Molecular Biology Intelligence Unit

The Nuclear Matrix and Spatial Organization of Chromosomal DNA Domains

Sergey V. Razin, Ph.D.

Institute of Gene Biology
Russian Academy of Sciences
Moscow, Russia

CHAPMAN & HALL
ITP An International Thomson Publishing Company

New York • Albany • Bonn • Boston • Cincinnati • Detroit • London • Madrid • Melbourne • Mexico City • Pacific Grove • Paris • San Francisco • Singapore • Tokyo • Toronto • Washington

R.G. Landes Company
Austin

Molecular Biology Intelligence Unit

THE NUCLEAR MATRIX AND SPATIAL ORGANIZATION OF CHROMOSOMAL DNA DOMAINS

R.G. LANDES COMPANY
Austin, Texas, U.S.A.

Printed in the U.S.A.

Please address all inquiries to the Publishers:
R.G. Landes Company, 810 S. Church Street, Georgetown, Texas, U.S.A. 78626
Phone: 512/ 863 7762; FAX: 512/ 863 0081

North American distributor:
Chapman & Hall, 115 Fifth Avenue, New York, New York, U.S.A. 10003

CHAPMAN & HALL

U.S. and Canada ISBN: 0-412-13371-7

While the authors, editors and publisher believe that drug selection and dosage and the specifications and usage of equipment and devices, as set forth in this book, are in accord with current recommendations and practice at the time of publication, they make no warranty, expressed or implied, with respect to material described in this book. In view of the ongoing research, equipment development, changes in governmental regulations and the rapid accumulation of information relating to the biomedical sciences, the reader is urged to carefully review and evaluate the information provided herein.

Library of Congress Cataloging-in-Publication Data
CIP applied for, but not received as of publication date.

Publisher's Note

R.G. Landes Company publishes six book series: *Medical Intelligence Unit, Molecular Biology Intelligence Unit, Neuroscience Intelligence Unit, Tissue Engineering Intelligence Unit, Biotechnology Intelligence Unit* and *Environmental Intelligence Unit.* The authors of our books are acknowledged leaders in their fields and the topics are unique. Almost without exception, no other similar books exist on these topics.

Our goal is to publish books in important and rapidly changing areas of bioscience and environment for sophisticated researchers and clinicians. To achieve this goal, we have accelerated our publishing program to conform to the fast pace in which information grows in bioscience. Most of our books are published within 90 to 120 days of receipt of the manuscript. We would like to thank our readers for their continuing interest and welcome any comments or suggestions they may have for future books.

Shyamali Ghosh
Publications Director
R.G. Landes Company

CONTENTS

PREFACE

More than 20 years ago it was demonstrated that eukaryotic DNA is organized into large (50-150 kb) loops by periodic attachment to the high salt insoluble nuclear matrix or scaffold. The specificity of DNA organization into loops has been intensively studied with the aim to find out whether such loops correspond to the functional units of the genome, such as replicons and transcriptions. These studies have resulted in controversial findings entailing conflicting conclusions. Nevertheless, most of the researchers agree on several items. Thus, it seems clear that the organization of chromosomal DNA into loops is not random. It seems equally clear that this organization is related to the functional organization of the genome. Furthermore, numerous evidence indicates that the interaction of DNA with the nuclear matrix is important for a proper organization of biosynthetic processes in eukaryotic cell nuclei.

The significance of a proper spatial organization of the eukaryotic genome as well as the significance of functional compartmentalization of eukaryotic cell nuclei were recognized long ago. At first this recognition followed from the logical analysis of problems that should be "solved" by cells of higher eukaryotes. Here are some of these problems: First, the about two meters long DNA molecule (actually a set of shorter molecules constituting individual chromosomes) must be packed somehow within the nuclei which typically have a diameter of about 10 microns. Second, in spite of a very compact packaging of DNA within eukaryotic cell nuclei, the transcribing regions should be accessible for RNA polymerases and the trans-acting factors. The patterns of transcribed regions are distinct in differentiated cells of various lineages. Hence, there should be a mechanism making almost any genomic region preferentially accessible for trans-acting factors and for the transcription machinery. Furthermore, there should be some pathways for effective transport of the products of transcription. The expressing genes (which again are not the same in differentiated cells of various lineages) should be located in the vicinity of the above pathways. Finally, in proliferating cells the whole DNA should replicate within a relatively short time interval (normally, within several hours). From the logical point of view it seemed obvious that to permit the diversity of functions the genome should be organized as a set of quasi-independent structural-functional domains. The discovery of chromosomal DNA loops fixed at the nuclear matrix constituted the first step toward the identification and experimental analysis of these domains.

A progress in any field of experimental studies depends on the availability of appropriate experimental tools. Recent years have been marked by an outburst of interest in the problem of spatial organization of the eukaryotic genome and by a fast progress in corresponding studies. This progress became possible with the elaboration of several new experimental approaches, such as confocal microscopy, multicolor fluorescence in situ hybridization (FISH), separation of large DNA molecules by pulsed-field-gel-electrophoresis and methods for creation of transgenic animals. Utilization of the above approaches made it possible to combine structural and functional studies of eukaryotic chromatin and chromosomes. This resulted in a new vision of an extremely complex structural-functional organization of eukaryotic cell nuclei.

The aim of this book is to discuss different questions concerning the spatial organization of the genome and its relation to the functional genome organization. However, it is clear that chromosomes occupy the major part of nuclei. It is not easy even in discussion (not to mention experiments) to separate interphase chromosomes from other nuclear components. Hence, it would be more correct to say that the whole problem of structural-functional organization of eukaryotic cell nuclei will be discussed with special emphasis on the genome spatial organization and functioning. The book contains twelve chapters that are logically interrelated. The nuclear matrix is a key structure which supports the functional compartmentalization of eukaryotic cell nuclei. That is why this book begins with the description of the nuclear matrix and the fate of its components in the course of mitosis (chapters 1 and 2). In the next eight chapters the organization of chromosomal DNA into large loops and the relation of this organization to the partitioning of the genome into functional units are discussed. The last two chapters present an attempt of synthesis of the present knowledge about the domain organization of the genome and about functional compartmentalization of eukaryotic cell nuclei. In the first of these concluding chapters (chapter 11) I discuss the rationales for supposing that chromosomal DNA loops constitute basic units of the genome organization and evolution. In the last chapter (chapter 12) different models describing the functional organization of eukaryotic cell nuclei are discussed.

As has already been mentioned at the beginning of this short introduction, the whole history of studies of the domain organization of eukaryotic genome is full of contradictions. Being directly involved in the above studies for the last 20 years I can hardly describe these studies as an indifferent observer. Although I have tried to present a balanced discussion of all vexed questions, some of my interpretations and conclusions may not be shared by other researchers. Furthermore, I have tried to make this book comprehensible for scientists and students who are not directly involved in studies of the structural-functional organization of the eukaryotic genome. Correspondingly, my discussion is concentrated on general problems. I am afraid that in some cases it is done at the expense of a more extended analysis of special questions. That is why each chapter is supplied with a short list of review articles written by the most eminent scientists working in the corresponding field. In these articles the reader may find more extended discussions of some special questions and also the argumentation for the conclusions which I do not share.

CHAPTER 1

What Is the Nuclear Matrix?

1.1. INTRODUCTION AND BRIEF HISTORICAL REVIEW

It has long been known that the main component of eukaryotic cell nuclei is DNA complexed with histones. Strange as it may seem, the elution of this component by high-salt extraction after mild nuclease treatment does not cause the disruption of nuclei. The residual (high-salt resistant) fraction was found to retain the shape and some morphological features of the nuclei, for example, a dense area identified as residual nucleoli and internal thin (~100 Å) filaments associated with ~200 Å granules (Fig. 1.1).[1,2] The filaments were proposed to be nuclear skeletal structures to which ribonucleoprotein granules, as well as deoxyribonucleoprotein fibrils of chromosomes, were attached (Fig. 1.1C).[1] Although initial observations of high-salt-insoluble residual nuclear structures were made in the early 1960s,[1,2] these structures became really known only after their comprehensive studies carried out by Berezney and Coffey more than ten years later.[3-5] Berezney and Coffey suggested the term "nuclear matrix" to call such residual nuclei. One of the key observations of these authors which drew the attention of scientists to the nuclear matrix was the demonstration of an association between active replication complexes and the nuclear matrix.[4] Berezney and Coffey proposed a protocol for nuclear matrix isolation which included treatment of isolated nuclei with nucleases (in experiments with rat liver nuclei this step may be omitted as both DNA and RNA are partially degraded by endogenous nucleases in the course of nuclei isolation), sequential extractions with low salt (0.2 mM $MgCl_2$, 10 mM Tris-HCl) and

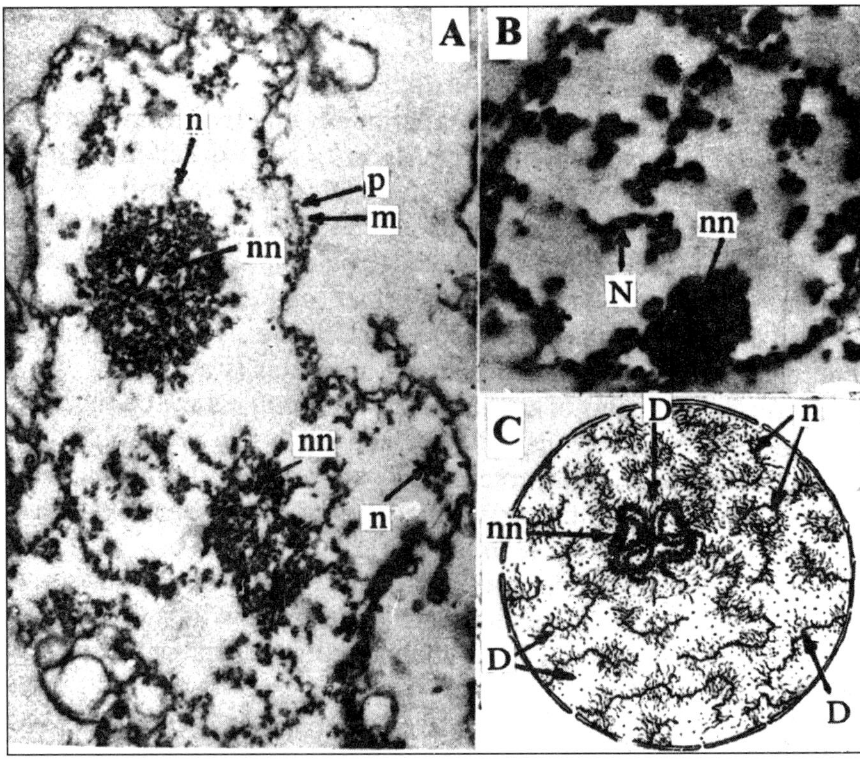

Fig. 1.1. Electron microscopy of the isolated rat liver nuclei. (A) Treatment with DNase I and then with O.14 M and 2 M NaCl; (B) Treatment with DNase I and then with 0.14 M NaCl; (C) The hypothetical model of cell nucleus based on the results obtained. n, Nucleonemas; D, DNP fibrils; nn, nucleolus; m, nuclear membrane with pores (p). N, thick fibrils formed after DNase I and 0.14 M NaCl treatment. This figure is combined from pictures taken with permission of authors from Georgiev GP, Chentsov YuS. Docladi Akademii Nauk SSSR (in Russian) 1960; 132: 199-202. Copyright retained by authors.

high salt (2M NaCl) solutions, extraction with 1% Triton X100 and, finally, the second treatment with nucleases.

Nuclear matrices isolated by the above protocol were composed of a nuclear lamina with residual nuclear pore complexes, residual nucleoli and a network of filaments connecting the residual nucleoli with the lamina (Fig. 1.2). Initial analysis of the nuclear matrix protein composition demonstrated lamins to be the major components.[3-5] Subsequent studies showed the protein composition of the nuclear matrix to be more complex.[6-15] Although according to the present data the nuclear matrix may contain more than a hundred of different proteins,[13,16] most of these proteins are either minor or (and) cell-type specific components.[13] Detailed analysis of the protein composition of the rat liver nuclear matrix made it possible to recognize approximately 12 major protein components.[15]

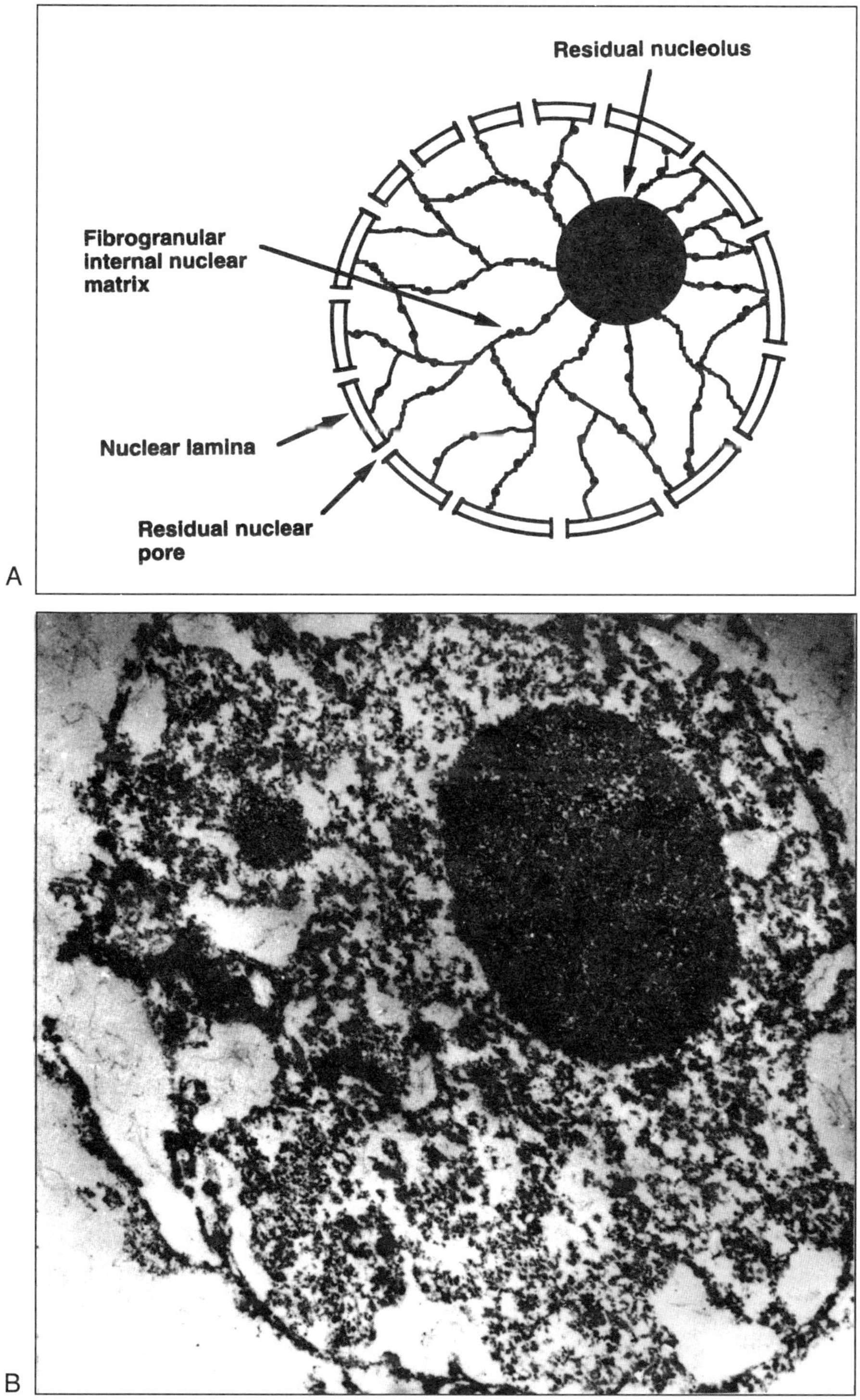

Fig. 1.2 (A) scheme of nuclear matrix. (B) An electron microscopic picture of the nuclear matrix from a chicken erythroblast cell.

Of these, 5 proteins were identified as lamins A, B and C, the nucleolar protein B-23 and the residual protein of nuclear RNP particles. The other eight proteins with estimated molecular weights of 125 kDa, 105 kDa, 60-75 kDa (four different basic proteins) and 42-48 kDa (two acidic proteins) differed from any of the previously characterized nuclear proteins. They were termed nuclear matrins.[15] Although four matrins are similar by their molecular weights to the lamins and can be separated from the latter only by two-dimensional gel electrophoresis, these matrins have a strictly interior nuclear localization as shown by subfractionation immunoblot experiments and by indirect immunofluorescence microscopy.[15] Further analysis has demonstrated that at least two of the 65-70 kDa matrins are DNA-binding proteins with predicted zinc finger motifs.[17]

As has been mentioned before, the nuclear matrix isolated by the sequential extraction procedure of Berezney and Coffey[4-6] can be subdivided into a nuclear lamina, a residual nucleolus and a so-called internal or diffuse matrix which is a fibrogranular network interconnecting the lamina with the residual nucleolus. The stability of these three components of the nuclear matrix differs significantly—the internal matrix being the most fragile part of the whole structure. No wonder that different deviations from the standard procedure resulted in preparation of nuclear matrices with differently preserved internal parts. In extreme cases (using thiol reagents in the course of extractions), it is possible to solubilize the whole internal matrix and to isolate an empty nuclear lamina, or nuclear shell.[10] This procedure constitutes a useful tool for separation of internal matrix proteins from lamins. Several types of treatment, such as incubation at 37°C[18-20] or exposure to oxidizing agents[10,20-24] were reported to stabilize the internal matrix.

It is not surprising that the protein composition of matrices isolated by different extraction procedures differs significantly.[10,18-26] That is why many scientists prefer to define the nuclear matrix operationally. For example, the nuclear matrix isolated according to the original protocol of Berezney and Coffey is defined as a subnuclear fraction which can be obtained by sequential extraction of isolated nuclei with low- and high-salt (2M NaCl) solutions followed by extraction with a nonionic detergent solution and an extensive treatment with DNase I and RNase A. Although this definition seems perfectly correct, it gives no idea of what the nuclear matrix is. It is, however, clear that nuclear matrices isolated from different sources by different procedures have many common features (see ref. 27 for a discussion and references). For example, almost all of them are composed of a lamina, a residual nucleolus and a more or less preserved internal matrix. Differences in the protein composition of matrices isolated by various procedures seem to be quantitative rather then qualitative.[18,20] Many components of the most elusive internal nuclear matrix have been located recently to the nuclear interior by immunofluorescence in situ hybridization, supporting the conclusion that nuclear matrices exist in nonextracted cells.[15,28] Furthermore, recent advances in the studies of nuclear compartmentalization (including compartmentalization of replication, transcription and splicing of primary transcripts), as well as the studies of the spatial organization of mRNA transport within nuclei, strongly support the idea that there should be some organizing nonchromatin internal structure in eukaryotic cell nuclei.[29-35] Hence, a simple operational definition of the nuclear matrix can hardly satisfy scientists presently working in this area. It seems more appropriate to give a more precise descriptive definition of the nuclear matrix. We see two possibilities to give such a definition and their discussion comes.

1.2. NUCLEAR MATRIX AS A SKELETAL STRUCTURE OF NUCLEI

Initially the matrix was considered as a supporting structure (skeleton or scaffold) which organizes the internal nuclear volume and provides surfaces for the attachment of multienzyme complexes involved in DNA and RNA metabolism.[35,36] This point of view was confirmed by the electron microscopic analysis carried out in the laboratory headed by Sheldon Penman. In this laboratory, a modified stepwise extraction procedure was developed with the aim to avoid a sharp increase in ionic strength of an extraction buffer. Application of this extraction procedure for isolation of nuclear matrices from different cells permitted visualizing of a fine network of 9 nm and 13 nm nuclear matrix filaments carrying some dense bodies (Fig. 1.3).[13,28,37,38] This filamentous network retained 70% of nuclear RNA (including both ribosomal RNA precursors and heterogeneous nuclear RNA). Treatment with RNase A removes the core filaments suggesting that RNA constitutes an integral part of these filaments. Direct extraction of chromatin by treatment of isolated nuclei with 2M NaCl solution does not uncover the core filaments of the above type. Instead, one can observe a diffuse reticulum of 30-50 Å fibrils which are not sensitive to RNase treatment.[3,5,39] It may as well happen that these fibrils represent RNase-sensitive core filaments covered by proteins. Evidence indicating that RNA may play an important role in supporting the nuclear matrix architecture was also presented by Bouvier et al.[40,41] According to these authors, the treatment of nuclei with RNase A prior to chromatin extraction by 2M NaCl solution results in the isolation of empty nuclear shells lacking an internal nuclear matrix. However, the same treatment has little effect if carried out after high-salt extraction.[40,41] Again, it was proposed that in the course of high-salt extraction the proteins covering RNA aggregated to form a filamentous network possibly stabilized by disulfide bridges.[40,41] An important role of RNA and RNP in supporting the nuclear matrix architecture has also been demonstrated in several other studies.[42-47] The question about the nature of proteins constituting the internal nuclear matrix remains controversial. According to the data of Berezney and collaborators, these may be **matrins** (see above and also refs. 15, 39). Some results suggest that **actin** may be a structural component of the internal nuclear matrix.[12,48-55] An important question is, however, whether actin detected in nuclear matrices appears as a result of contamination of preparations with cytoplasmic actin. Several observations argue against this possibility. The nuclear matrix actin was reported to contain acidic species that were scarcely detectable in cytoplasm.[52] Fibrogranular structures of the internal nuclear matrix can be stained with anti-actin antibodies.[50]

As far as a possible structural function of actin in the nuclear matrix is concerned, it is important to mention that experimental procedures known to depolymerize F-actin cause a release from the nuclear matrix of a major portion of RNA and of some matrix-bound proteins.[51,53,54] Hence, it may happen that actin covers and stabilizes the RNase-sensitive core filaments described by Penman and collaborators.

Some recent data indicate that **lamins** not only form the peripheral nuclear lamina but also participate in the construction of the internal nuclear matrix.[56-58] According to the above cited authors, in intact nuclei the internal lamins (or lamin filaments) are covered by huge masses of chromatin and hence cannot be stained

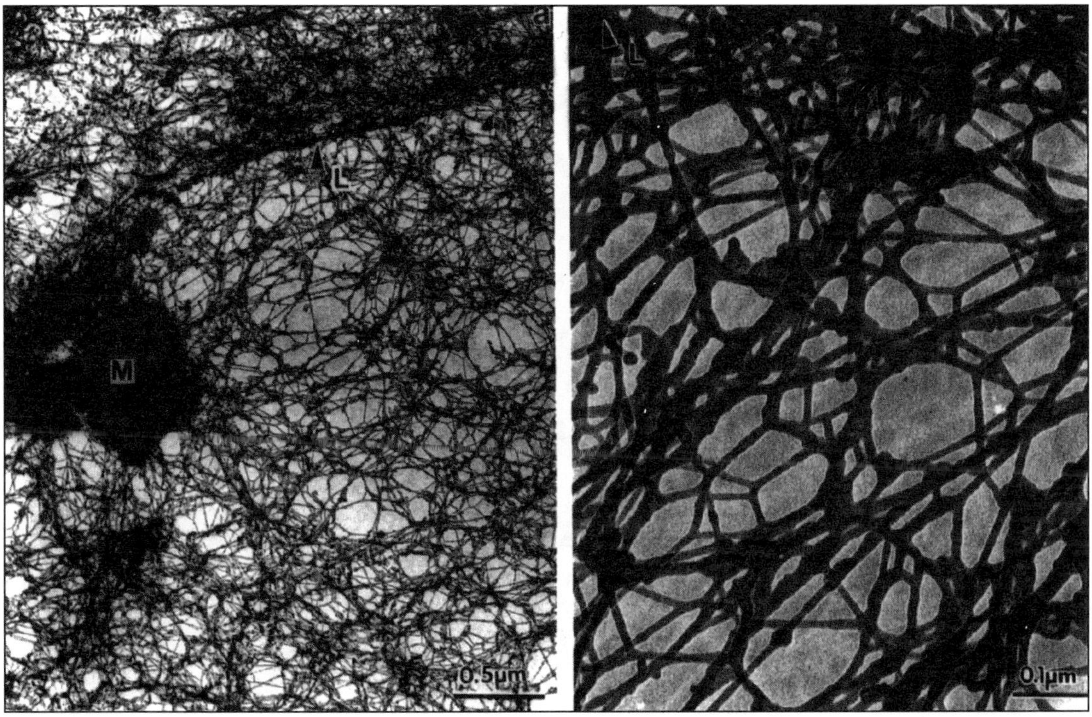

Fig. 1.3. The core filaments of the MCF-7 nuclear matrix. (a) Core filaments. The nuclear lamina (L) connected the network of nuclear core filaments (below) to the cytoplasmic intermediate filament network (above). A residual mass (M) remained enmeshed in the network of filaments. (b) Higher order view of core filaments. In this higher magnification view, the tripartite junctions between core filaments could be seen more clearly. The network of core filaments is connected to the nuclear lamina (L). (Reprinted with permission of The Rockefeller University Press from He DC, Nickerson JA, Penman S. J Cell Biol 1990; 110:569-580. © 1990 by The Rockefeller University Press).

with antibodies. However, they become accessible to antibodies after extraction of chromatin (i.e., in nuclear matrices). It is known that lamins are closely related to cytoplasmic intermediate filament proteins.[59-61] In this connection, it is worth mentioning that intermediate-type filaments (10 nm wide with an axial 23 nm repeat) were observed in residual nuclei depleted of chromatin by electroelution in physiological ionic strength[62] (Fig. 1.4). Intermediate filament-related components were also identified in plant cell nuclei by immunostaining with antibodies raised against the nuclear matrix proteins.[63] It should be mentioned that lamins were detected in nuclear speckles, including the replication foci.[56,57] Hence, in addition to supporting the nuclear envelope and participating in the formation of internal nuclear matrix filaments, lamins may play some other functions in nuclear space compartmentalization.

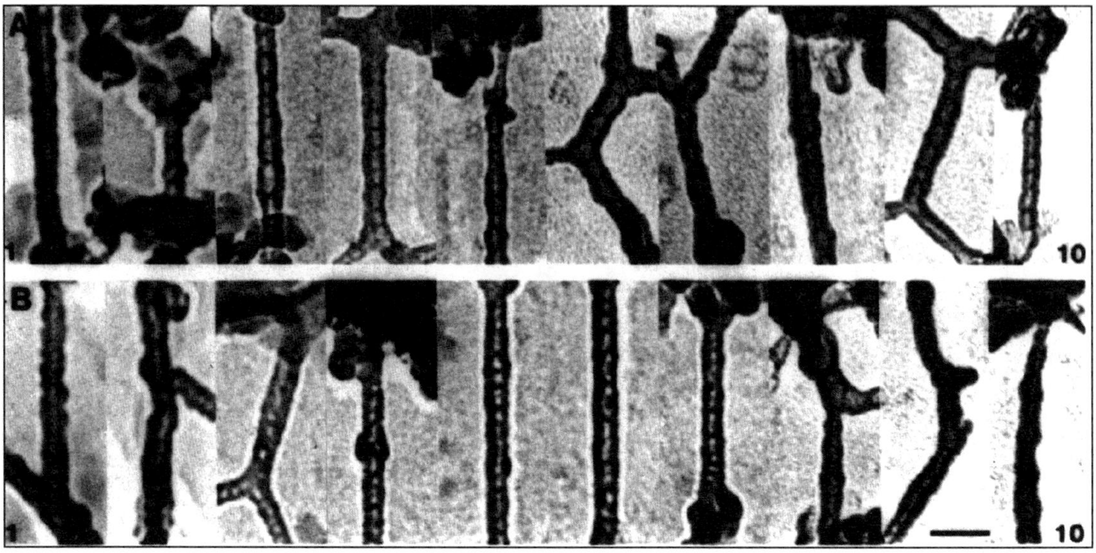

Fig. 1.4. Electron micrographs of 10 typical nuclear (A) and cytoplasmic (B) filaments. A1-2 and B1-2: low-angle (20°) carbon coating. A3-5 and B3-7: high angle (70°) rotary shadowing. A6-10 and B8-10: low angle rotary platinum/palladium shadowing followed by high angle rotary carbon shadowing. The bar represents 50 nm. (Reprinted with permission of IRL Press Ltd. from Jackson DA, Cook PR. EMBO J 1988; 7:3667-3677. © 1988 EMBO/IRL Press Ltd.).

The list of proteins assumed to participate in the construction of the nuclear skeleton can be extended. Having no intention to present a full roll of candidates, we shall mention here only **DNA topoisomerase II**[64] and the nuclear mitotic apparatus protein (**NuMA**)[65-67] which will be discussed in more detail in the next chapters.

Although an exact composition of internal nuclear matrix filaments remains to be determined, the bare fact of the existence of such filaments can hardly be questioned. However, it is difficult (if at all possible) to isolate these filaments as such. In nuclear matrices they are interconnected with a number of proteins, such as proteins involved in DNA replication, transcription, splicing and transport of messenger RNA. In order to avoid confusion we suggest to use the term **nuclear skeleton** for designation of the nuclear lamina and the network of filaments supporting the internal nuclear architecture. Many proteins involved in the formation of nuclear pore complexes cannot be separated from the nuclear lamina by 2M NaCl extraction. Hence, these proteins may also be considered as components of the nuclear skeleton. Detailed description of the organization of nuclear pores is beyond the scope of this chapter (for a review see ref. 68).

The nuclear skeleton is essential for maintaining both the shape and the internal structural organization of eukaryotic cell nuclei. Hence it is not surprising that proteolytic cleavage of major components of the nuclear skeleton (such as lamin B or NuMA protein) is one of the earliest events in the course of apoptotic degradation of nuclei.[69] In human thymocytes triggered to undergo apoptosis by dexametasone treatment, the cleavage of lamin B and NuMA protein was observed within 30 minutes after the treatment, i.e., before any DNA fragmentation could be detected.[69]

Clearly, the nuclear skeleton constitutes a part of the nuclear matrix, the latter being a more complex structure. It is also obvious that the system of filaments supporting an architecture of a whole cell is unified, i.e., that the nuclear skeleton is integrated into the network of cytoskeletal filaments. This interconnection between the nuclear skeleton and the cytoskeleton is clearly seen on electron micrographs made by Penman and collaborators.[28,37,38] It has been proposed that beside supporting cellular architecture the integrated cytoskeletal and nuclear matrix networks also serve for transduction of extracellular signals to nuclei and, after all, to DNA.[70]

1.3. NUCLEAR MATRIX AS A SPECIAL NUCLEAR COMPARTMENT

From the present knowledge about the distribution of proteins between the nuclear matrix and a combined extract it seems reasonable to define the nuclear matrix as **a high-salt-insoluble nuclear compartment**. This compartment will include exclusively matrix proteins (for example, nuclear skeleton proteins) and also proteins that are permanently or temporarily deposited in the nuclear matrix. The deposition of different proteins in the nuclear matrix may be related to some posttranslational modification of these proteins and may constitute a part of regulation cascades. To illustrate this possibility, we may focus our attention on a tumor suppressor retinoblastoma (p105Rb) protein. Hypophosphorylated p105Rb is tightly bound to the nuclear matrix.[71] The same is true of several p105Rb-binding viral oncoproteins.[71] However, the hyperphosphorylated form of p105Rb, as well as a mutated form of this protein which does not possess the ability to prevent cells from entering the next cycle, is not associated with the nuclear matrix. Another potential tumor suppressor protein, PML, behaves similarly to p105Rb.[72]

To give another example of the important role of the three dimensional organization of cell nuclei in the regulation of functional processes, we can refer to the regulation of transcription by trans-acting factors. It is becoming increasingly evident that some transcriptional factors are deposited in the nuclear matrix compartment in a cell-type-specific and development stage-specific fashion. In some cases, the regulatory DNA sequences become also detained in the matrix compartment due to the interaction with the above proteins[73-77] (for a review see refs. 78, 79).

Cell-cycle regulators (not only p105Rb protein, but also cdk2 and cdc2 kinases[80]) and transcription factors are only two examples of large groups of nuclear proteins which are partitioned between the high-salt-insoluble (nuclear matrix) and soluble compartments of eukaryotic cell nuclei. Other groups of proteins with a similar dual distribution are enumerated below. These are:

1. **Enzymes involved in DNA replication and postreplicative modifications**, including DNA polymerase α, DNA primase, 3'-5' endonuclease, RNase H, DNA methylase and proliferating cell nuclear antigen (PCNA).[81-89]
2. **Enzymes involved in transcription and splicing of messenger RNA and proteins of nuclear RNP-particles.** This group includes RNA-polymerase II[90,91] and the above discussed transcription factors. Furthermore, many proteins involved in splicing of mRNA were reported to be components of the nuclear matrix.[92-97] One of the proteins ensuring the integration of the splicing machinery into the nuclear matrix seems to be NuMA which apparently fulfills multiple functions in interphase nuclei and during mitosis.[66,67] The proteins of RNP core particles (along with RNA itself) are even considered to be a part of the nuclear skeleton, i.e., they are supposed to play a structural role supporting the nuclear architecture.[25,98-102] It should be mentioned that similarly to DNA replication, the processes of transcription and splicing were shown to occur within specific foci or speckles.[31,33,34,103] It is possible that the principal

function of the nuclear matrix proteins is to form these foci and hold there all components of the transcription and splicing machinery.

3. **Proteins of the mitotic apparatus.** This group includes several well-characterized proteins. These are CENP-B, CENP-C, CENP-F and NuMA. CENP-B and CENP-C are constitutive components of kinetochores which are present at centromeres both in mitosis and in interphase. CENP-B and CENP-C cannot be eluted from nuclei by the high salt extraction[104] and hence may be considered as components of the nuclear matrix. CENP-F seems to be simply deposited in the nuclear matrix compartment during interphase. During mitosis this protein is associated with kinetochores from prometaphase till early anaphase. In anaphase it moves to the spindle midzone and is finally (in telophase) concentrated within the intracellular bridge.[105,106] NuMA has already been mentioned several times in this chapter. It is a long coiled-coil protein which is likely to constitute in interphase an integral structural part of the nuclear skeleton.[66,67] During mitosis NuMA is associated with the mitotic spindle (predominantly with the pericentriolar region). NuMA plays an essential role in mitosis since its inactivation by antibodies causes mitotic arrest and subsequent formation of micronuclei (for a review see ref. 107). From our present knowledge, it would be appropriate to say that NuMA participates in either organization or stabilization of the mitotic spindle. It seems possible that in interphase this protein plays a similar role in respect to nuclear matrix filaments.
4. **DNA-binding proteins** (including those which bind DNA with a strict or limited sequence specificity, but which are not transcriptional factors). These proteins possibly participate in the anchorage of chromosomal DNA loops to the nuclear matrix and will be discussed in detail in chapter 6 of the present manuscript.
5. **Regulators of the cell cycle.** Some of the nuclear matrix proteins involved (or possibly involved) in the regulation of the cell cycle have already been mentioned before. These are p105Rb with associated proteins,

PML, cdk2 and cdc2. Recently it has been shown that casein kinase 2 is present in purified nuclear matrices.[108] Different cellular and viral oncoproteins were reported to reside completely or partially (in some cases depending on the stage of the cell cycle) in the nuclear matrix compartment. These are c-myc, L-myc, N-myc, and v-myc oncoproteins,[109-113] c-myb and v-myb oncoproteins,[54,114-118] E7 protein of human papillomavirus 16,[119,120] simian virus 40 T antigen[121] and adenovirus 2 E1a protein.[122]

6. **Proteins involved in the control of chromatin dynamics.** Two key enzymes controlling the level of histone acetyllation, namely histoneacetyltransferase and histone diacetyllase, were found to reside in the internal nuclear matrix.[123-125] Another nuclear matrix component possibly involved in the control of chromatin dynamics is the p62/calmodulin complex which is thought to participate in the organization of heterochromatin.[126]
7. **Receptors and receptor-binding proteins.** A number of hormone receptors and their nuclear binding sites were found to reside in the nuclear matrix. The list includes estrogen receptors[127] and their acceptor sites,[128] androgen receptors[127] and their binding sites,[129,130] mineralocorticoid receptors,[131] glucocorticoid receptors,[132] and progesterone receptor acceptor sites.[133,134] Most of the above receptors and receptor-binding sites were found to reside in the internal matrix.[127,135-137]

Some nuclear matrix proteins cannot be included in any of the above listed groups. As it becomes clear that all or almost all functional processes in nuclei occur in the nuclear matrix compartment, one may expect that the number of characterized nuclear matrix proteins will rapidly increase. As has been mentioned in the beginning of this chapter, the spectrum of nuclear matrix proteins depends on the cell lineage. Furthermore, it changes in the course of cell differentiation and transformation. A possibility to use some nuclear matrix proteins as markers for cancer diagnosis is seriously considered (for a review see refs. 138, 139).

1.4. CONCLUSIONS

We suggest to define the nuclear matrix as **a high-salt-insoluble nuclear compartment which retains the shape and some morphological features of the nuclei.** At the same time, we suggest to use the term "**nuclear skeleton**" for designation of the supporting structure (scaffold) of the nuclear matrix. Although we shall try to use these definitions throughout the whole manuscript, it will be useful also to remember several other terms that can frequently be found in special literature. These terms are explained below:

- **nuclear cage, nuclear ghost, nucleoid core** are the old terms used in early works of P. Cook and collaborators[36,89,140] for designation of nuclear remnants in cells extracted with nonionic detergent and 2M NaCl solution.
- **in situ nuclear matrix** may also be called cellular matrix as it contains the nuclear matrix immersed into the cytoskeleton and attached to a substratum (for example, to a glass coverslip) through the cytoskeleton filaments. In situ nuclear matrices can be prepared by sequential extractions of cells attached to a substratum and their treatment with nucleases.[141]
- **nucleoskeleton** is a term suggested by P. Cook and collaborators[62,142,143] for designation of nuclear remnants from which chromatin was fully or partially removed by electroelution in physiological ionic strength.
- **nuclear scaffold** is a term suggested by U. Laemmli and collaborators[144] for designation of nuclear remnants from which chromatin was removed by extraction with a mild ionic detergent, lithium diiodosalicylate (LIS).

RELATED REVIEW ARTICLES

Bouteille M, Bouvier D, Seve AP. Heterogeneity and territorial organization of the nuclear matrix and related structures. Int Rev Cytol 1983; 83:135-182.

Kaufmann SH, Fields AP, Shaper JH. The nuclear matrix: current concepts and unanswered questions. Methods Achiev Exp Pathol 1986; 12:41-171.

Verheijen R, van Venrooij W Ramaekers F. The nuclear matrix: structure and composition. J Cell Sci 1988; 90:11-36.

Fisher PA. Chromosomes and chromatin structure: the extrachromosomal karyoskeleton. Curr Opin Cell Biol 1989; 1:447-453.

Van Driel R, Humbel B, De Jong L. The nucleus: a black box being opened. J Cell Biochem 1991; 47:311-316.

Berezney R. The nuclear matrix: a heuristic model for investigating genomic organization and function in the cell nucleus. J Cell Biochem 1991; 47:109-123.

Fey EG, Bangs P, Sparks C et al. The nuclear matrix: defining structural and functional roles. Crit Rev Eukaryot Gene Exp 1991; 1:127-143.

Jack RS, Eggert H. The elusive nuclear matrix. Eur J Biochem 1992; 209:503-509.

Nickerson JA, Penman S. Localization of nuclear matrix core filament proteins at interphase and mitosis. Cell Biol Int Rep 1992; 16:811-826.

Stuurman N, De Jong L, Van Driel R. Nuclear frameworks: concepts and operational definitions. Cell Biol Int Rep 1992; 16:837-852.

Paulin-Levasseur M. Intermediate filament proteins: many unanswered questions. few unquestioned answers. Biochem Cell Biol 1992; 70:842-848.

Penman S. Rethinking cell structure. Proc Natl Acad Sci USA 1995; 92:5251-5257.

Berezney R, Mortillaro MJ, Ma H et al. The nuclear matrix: A structural milieu for nuclear genomic function. Int Rev Cytol 1995; 162A:1-65.

References

1. Georgiev GP, Chentsov YuS. On the structure of cell nuclei: experimental electron microscopic investigation of isolated nuclei. Docladi Akademii Nauk SSSR (in Russian) 1960; 132:199-202.
2. Zbarsky IB, Dmitrieva NP, Yermolaeva LP. On the structure of tumour cell nuclei. Exp Cell Res 1962; 27:573-576.
3. Berezney R, Coffey DS. Identification of a nuclear protein matrix. Biochem Biophys Res Commun 1974; 60:410-1417.
4. Berezney R, Coffey DS. Nuclear protein matrix: association with newly synthesized DNA. Science 1975; 189:291-292.
5. Berezney R, Coffey DS. Nuclear matrix: Isolation and characterization of a framework structure from rat liver nuclei. J Cell Biol 1977; 73:616-637.
6. Berezney R. Effect of protease inhibitors on matrix proteins and the association of replicating DNA. Exp Cell Res 1979; 123:411-414.
7. Peters KE, Comings DE. Two-dimensional gel electrophoresis of rat liver nuclear washes, nuclear matrix and hnRNA proteins. J Cell Biol 1980; 86:135-155.
8. Peters KF, Okada TA, Comings DE. Chinese hamster nuclear proteins. An electrophoretic analysis of interphase, metaphase and nuclear matrix preparations. Eur J Biochem 1982; 129:221-232.

9. Peick AC, van der Velden HM, Rijken AA et al. Protein composition of the chromosomal scaffold and interphase nuclear matrix. Chromosoma 1985; 91:137-144.
10. Kaufmann SH, Coffey DS, Shaper JH. Considerations in the isolation of rat liver nuclear matrix, nuclear envelope, and pore complex lamina. Exp Cell Res 1981; 132:105-123.
11. Razin SV, Chernokhvostov VV, Roodin AV et al. Proteins tightly bound to DNA in the regions of DNA attachment to the skeletal structures of interphase nuclei and metaphase chromosomes. Cell 1981; 27:65-73.
12. Capco DG, Wan KM, Penman S. The nuclear matrix: three-dimensional architecture and protein composition. Cell 1982; 29:847-858.
13 Fey EG, Penman S. Nuclear matrix proteins reflect cell type of origin in cultured human cells. Proc Natl Acad Sci USA 1988, 85:21 125.
14. Cupo JF. Electrophoretic analysis of nuclear matrix proteins and the potential clinical applications. J Chromatogr 1991; 569:389-406.
15. Nakayasu H, Berezney R. Nuclear matrins: identification of the major nuclear matrix proteins. Proc Natl Acad Sci USA 1991; 88:10312-10316.
16. Dell'Orco RT, Whittle WL. Nuclear matrix composition and in vitro cellular senescence. Exp Gerontol 1994; 29:39-149.
17. Hakes DJ, Berezney R. DNA binding properties of the nuclear matrix and individual nuclear matrix proteins. Evidence for salt-resistant DNA binding sites. J Biol Chem 1991; 266:11131-11140.
18. Belgrader P, Siegel AJ, Berezney R. A comprehensive study on the isolation and characterization of the HeLa S3 nuclear matrix. J Cell Sci 1991; 98:281-291.
19. Martelli AM, Gilmour RS, Bareggi R et al. the effect of in vitro heat exposure on the recovery of nuclear matrix-bound DNA polymerase α activity during the different phases of the cell cycle in synchronized HeLa S3 cells. Exp Cell Res 1992; 201:470-476.
20. Martelli AM, Manzoli L, Rubbini S et al. The protein composition of Friend cell nuclear matrix stabilized by various treatments. Different recovery of nucleolar proteins B23 and C23 and nuclear lamins. Biol Cell 1995; 83:15-22.
21. Kaufmann SH, Shapper JH. A subset of non-histone nuclear proteins reversibly stabilized by the sulfhydril cross-linking reagent tetrathionate. Exp Cell Res 1984; 155:477-495.
22. Kaufmann SH, Shaper JH. Association of topoisomerase II with the hepatoma cell nuclear matrix: The role of intermolecular disulfide bond formation. Exp Cell Res 1991; 192:511-523.
23. Kaufmann SH, Brunet G, Talbot B et al. Association of poly(ADP-ribose) polymerase with the nuclear matrix: The rele of intermolecular disulfide bond formation, RNA retention, and cell type. Exp Cell Res 1991; 192:524-535.

24. Stuurman N, Floore A, Colen A et al. Stabilization of the nuclear matrix by disulfide bridges: Identification of matrix polypeptides that form disulfides. Exp Cell Res 1992; 200:285-294.
25. Verheijen R, Kuijpers H, Vooijs P et al. Protein composition of nuclear matrix preparations from HeLa cells: an immunochemical approach. J Cell Sci 1986; 80:103-122.
26. Kallajoki M, Osborn M. Gel electrophoresis of nuclear matrix fractions isolated from different human cell lines. Electrophoresis 1994; 15:520-528.
27. Berezney R. Organization and functions of the nuclear matrix. In: Hnilica LS, ed. Chromosomal Nonhistone Proteins. Boca Raton, FL: CRC Press, 1984; 4:119-180.
28. Penman S. Rethinking cell structure. Proc Natl Acad Sci USA 1995; 92:5251-5257.
29. Spector DL. Higher order nuclear organization: Three-dimensional distribution of small nuclear ribonucleoprotein particles. Proc Natl Acad Sci USA 1990; 87:147-151.
30. Hoffman M. 1992. New evidence found for a nuclear matrix. Science 1992; 255:34.
31. Carter CC, Bowman D, Carrington, W et al. A three-dimensional view of precursor messenger RNA metabolism within mammalian nucleus. Science 1993; 259:1330-1335.
32. Lawrence JB, Carter KS, Xing X. Probing functional organization within the nucleus: is genome structure integrated with RNA metabolism. Cold Spring Harb Symp Quant Biol 1993; 58:807-818.
33. Xing Y, Johnson CV, Dobner PR et al. Higher level organization of individual gene transcription and RNA splicing. Science 1993; 259:326-1330.
34. Xing Y, Johnson CV, Moen Jr PT et al. Nonrandom gene organization: Structural arrangements of specific pre-mRNA transcription and splicing with SC-35 domains. J Cell Biol 1995; 131:1635-1647.
35. Baskin Y. Mapping the cell's nucleus. Science 1995; 268:1564-1565.
36. Jackson DA, McCready SJ, Cook PR. RNA is synthesized at the nuclear cage. Nature 1981; 292:552-555.
37. He DC, Nickerson JA, Penman S. Core filaments of the nuclear matrix. J Cell Biol 1990; 110:569-580.
38. Nickerson JA, Penman S. Localization of nuclear matrix core filament proteins at interphase and mitosis. Cell Biol Int Rep 1992; 16:811-826.
39. Berezney R. Matricin fibrils: isolation of the primary nuclear matrix polypeptides. Cell Biol Int Rep 1980; 4:167-173.
40. Bouvier D, Hubert J, Seve A-P et al. RNA is responsible for the three-dimensional organization of nuclear matrix proteins in HeLa cells. Biol Cell 1982; 43:143-146.
41. Bouvier D, Hubert J, Seve A-P et al. Structural aspects of intranuclear matrix desintegration upon RNase digestion of HeLa cell nuclei. Eur J Cell Biol 1985; 36:323-333.

42. Miller TE, Huang CY, Pogo AO. Rat liver nuclear skeleton and ribonucleoprotein complexes containing hnRNA. J Cell Biol 1978; 76:675-691.
43. Herman R, Weymouth L, Penman S. Heterogeneous nuclear RNA-protein fibers in chromatin-depleted nuclei. J Cell Biol 1978; 78:663-674.
44. Herlan G, Eckert WA, Kaffenberger W et al. Isolation and characterization of an RNA-containing nuclear matrix from tetrahymena macronuclei. Biochemistry 1979; 18:1782-1788.
45. Harris SG, Smith HC. SnRNP core protein enrichment in the nuclear matrix. Biochem Biophys Res Commun 1988; 152: 1383-1387.
46. Van Eekelen CAG, van Venrooij WJ. hn RNA and its attachment to a nuclear protein matrix. J Cell Biol 1981; 88:554-563.
47. Gallinaro H, Puvion E, Kisler L et al. Nuclear matrix and hnRNP share a common structural constituent associated with premessenger RNA. EMBO J 1983; 2:953-960.
48. Nakayasu H, Ueda K. Association of actin with the nuclear matrix from bovine lymphocytes. Exp Cell Res 1983; 143:55-62.
49. Nakayasu H, Ueda K. Small nuclear RNA-protein complex anchors on the actin filaments in bovine lymphocyte nuclear matrix. Cell Struct Funct (Japan) 1984; 9:317-325.
50. Nakayasu H, Ueda K. Ultrastructural localization of actin in nuclear matrices from mouse leukemia L5178Y cells. Cell Struct Funct (Japan) 1985; 10:305-309.
51. Nakayasu H, Ueda K. Association of rapidly-labeled RNAs with actin in nuclear matrix from mouse L5178Y cells. Exp Cell Res 1985; 60:319-330.
52. Nakayasu H, Ueda K. Preferential association of acidic actin with nuclei and nuclear matrix from mouse leukemia L5178Y cells. Exp Cell Res 1986; 163:327-336.
53. Schroder HC, Trolltsch D, Wenger R et al. Cytocholasin B selectively releases ovalbumin mRNA precursors but not the mature ovalbumin mRNA from hen oviduct nuclear matrix. Eur J Biochem 1987; 167:239-245.
54. Boyle WJ, Baluda MA. Subnuclear associations of the v-myb oncogene product and actin are dependent on ionic strength during nuclear isolation. Mol Cell Biol 1987; 7:3345-3348.
55. Crowley KS. Does the interchromatin compartment contain actin. Cell Biol Int Rep 1987; 7:537-546.
56. Bridger JM, Kill IR, O'Farrell M et al. Internal lamin structures within G1 nuclei of human dermal fibroblast. J Cell Sci 1993; 104:297-306.
57. Moir RD, Montag-Lowy M, Goldman RD. 1994. Dynamic properties of nuclear lamins: lamin B is associated with sites of DNA replication. J Cell Biol 1994; 125:1201-1212.

58. Hozak P, Sasseville AM-J, Raymond Y et al. Lamin proteins form an internal nucleoskeleton as well as peripheral lamina in human cells. J Cell Sci 1995; 108:635-644.
59. Fisher DZ, Chaudhary N, Blobel B. cDNA sequencing of nuclear lamins A and C reveals primary and secondary structural homology to intermediate filament proteins. Proc Natl Acad Sci USA 1986; 83:6450-6464.
60. McKeon FD, Kirschner MW, Caput D. Homologies in both primary and secondary structure between nuclear envelope and intermediate filament proteins. Nature 1986; 319:463-468.
61. Parry DAD, Steinert PM. Intermediate filament structure. Curr Opin Cell Biol 1992; 4:94-98.
62. Jackson DA, Cook PR. Visualization of a filamentous nucleoskeleton with a 23 nm axial repeat. EMBO J 1988; 7:3667-3677.
63. Beven A, Guan Y, Peart J et al. Monoclonal antibodies to plant nuclear matrix reveal intermediate filament-related components within the nucleus. J Cell Sci 1991; 98:293-302.
64. Berrious M, Osheroff N, Fisher P. In situ localization of DNA topoisomerase II, a major polypeptide of *Drosophila* nuclear matrix. Proc Natl Acad Sci USA 1985; 82:4142-4146.
65. Kallajoki M, Weber K, Osborn M. A 210 kd nuclear matrix protein is a functional part of the mitotic spindle: A microinjection study using SPN monoclonal antibodies. EMBO J 1991; 10: 3351-3362.
66. Zeng C, He D, Brinkley BR. Localization of NuMA protein isoforms in the nuclear matrix of mammalian cells. Cell Motil Cytoskeleton 1994; 29:167-176.
67. Zeng C, He D, Berget SM et al. Nuclear-mitotic apparatus protein: a structural interphase between the nucleoskeleton and RNA splicing. Proc Natl Acad Sci USA 1994; 91:1505-1509.
68. Pante N, Aebi U. Toward molecular understanding of the structure and function of the nuclear pore complex. Int Rev Cytol 1995; 162B:225-255.
69. Weaver VM, Carson CE, Walker PR et al. Degradation of nuclear matrix and DNA cleavage in apoptotic thymocytes. J Cell Sci 1996; 109:45-56.
70. Pienta KJ, Getzenberg RH, Coffey DS. Cell structure and DNA organization. Crit Rev Eukar Gene Exp 1991; 4:355-385.
71. Mancini MA, Shan B, Nickerson JA et al. The retinoblastoma gene product is a cell cycle dependent nuclear matrix-associated protein. Proc Natl Acad Sci USA 1994; 91:418-422.
72. Chang KS, Fan YH, Andreeff M et al. The PML gene encodes a phosphoprotein associated with the nuclear matrix. Blood 1995; 85:3646-3653.
73. Van Wijnen AJ, Bidwell JP, Fey EG et al. Nuclear matrix association of multiple sequence-specific DNA binding activities related

to SP1, ATF, CCAAT, C/ERB, OCT-1 and AP-1. Biochemistry 1993; 32:8397-83402.
74. Bidwell JP, van Wijnen AJ, Fey EG et al. Osteocalcin gene promoter-binding factors are tissue-specific nuclear matrix components. Proc Natl Acad Sci USA 1993; 90:3162-3166.
75. Vassetzky YS, De Moura Gallo CV, Bogdanova AN et al. The sequence-specific nuclear matrix-binding factor F6 is a chicken GATA-like protein. Mol Gen Genet 1993; 238:309-314.
76. Sun JM, Chen HY, Davie JR. Nuclear factor 1 is a component of the nuclear matrix. J Cell Biochem 1994; 55:252-263.
77. Bagchi M, Van Wijnen A, Katar M et al. Sequence-specific DNA binding activities of nuclear matrix proteins of mammalian lens epithelial cells. J Cell Biochem 1995; 58:1-5.
78. Getzenberg RH. Nuclear matrix and the regulation of gene expression: tissue specificity. J Cell Biochem 1994; 55:22-31.
79. Stein GS, van Wijnen AJ, Stein JL et al. Nuclear architecture supports integration of physiological regulatory signals for transcription of cell growth and tissue-specific genes during osteoblast differentiation. J Cell Biochem 1994; 55:4-15.
80. Jaumot M, Grana X, Giordano A et al. Cyclin/cdk2 complexes in the nuclear matrix of HeLa cells. Biochem Biophys Res Commun 1994; 203:1527-1534.
81. Smith HC, Berezney R. DNA polymerase alpha is tightly bound to the nuclear matrix of actively regenerating liver. Biochem Biophys Res Commun 1980; 97:1541-1547.
82. Smith HC, Berezney R. Dynamic domains of DNA polymerase alpha in regenerating rat liver. Biochemistry 1983; 22:3042-3046.
83. Foster KA, Collins JM. The interrelation between DNA synthesis rates and DNA polymerases bound to the nuclear matrix in synchronized HeLa cells. J Biol Chem 1985; 260:4229-4235.
84. Wood SH, Collins JM. Preferential binding of DNA primase to the nuclear matrix in HeLa cells. J Biol Chem 1986; 261: 7119-7122.
85. Tubo RA, Berezney R. Nuclear matrix bound DNA primase. J Biol Chem 1987; 262:6637-6642.
86. Hirose F, Yamamoto S, Yamaguchi M et al. Identification and subcellular localization of the polypeptide for chick DNA primase with a specific monoclonal antibody. J Biol Chem 1988; 263:2925-2933.
87. Tubo RA, Berezney R. Pre-replicative association of multiple replicative enzyme activities with the nuclear matrix during rat liver regeneration. J Biol Chem 1987; 262:1148-1154.
88. Tubo RA, Berezney R. Identification of 100 and 150 S DNA polymerase alpha-primase megacomplexes solubilized from the nuclear matrix of regenerating rat liver. J Biol Chem 1987; 262:5857-5865.
89. Bravo R, Macdonald-Bravo H. Existence of two populations of cyclin/proliferating cell nuclear antigen during the cell cycle: Asso-

ciation with DNA replication sites. J Cell Biol 1987; 105: 1549-1554.

90. Jackson DA, McCready SJ, Cook PR. Replication and transcription depend on attachment of DNA to the nuclear cage. J Cell Sci 1984; Suppl. 1:59-79.
91. Razin SV, Yarovaya OV. Initiated complexes of RNA polymerase II are concentrated in the nuclear skeleton associated DNA. Exp Cell Res 1985; 158:273-275.
92. Reuter R, Appel B, Bringmann P et al. 5'-Terminal caps of snRNAs are reactive with antibodies specific for 2,2,7-trimethylguanosine in whole cells and nuclear matrices. Double-label immunofluorescent studies with anti-m^3G antibodies and with anti-RNP and anti-Sm autoantibodies. Exp Cell Res 1984; 154:548-560.
93. Habets WJ, Berden JH, Hoch SO et al. Further characterization of Sm and U1 ribonucleoprotein antigens. Eur J Immunol 1985; 15:992-997.
94. Verheijen R, Kuijpers H, Vooijs P et al. Distribution of the 70k U1 RNA associated protein during interphase and mitosis. Correlation with other U RNP particles and proteins of the nuclear matrix. J Cell Sci 1986; 86:173-190.
95. Harris SG, Smith HC. SnRNP core protein enrichment in the nuclear matrix. Biochem Biophys Res Commun 1988; 152: 1383-1387.
96. Shen WC, Selvakumar D, Stanford DR et al. The Saccharomyces cerevisiae LOS1 gene involved in pre-tRNA splicing encodes a nuclear protein that behaves as a component of the nuclear matrix. 1993; 268:19436-19444.
97. Blencowe BJ, Nickerson JA, Issner R et al. Association of nuclear matrix antigens with exon-containing splicing complexes. J Cell Biol 1994; 127:593-607.
98. Fritzler MJ, Ali R, Tam EM. Antibodies from patients with mixed connective tissue disease react with heterogeneous nuclear ribonucleoprotein or ribonucleic acid (hnRNP/RNA) of the nuclear matrix. J Immunol 1984; 132:1216-1222.
99. Fey EG, Krochmalnic G, Penman S. The nonchromatin substructures of the nucleus: the ribonucleoprotein (RNP)-containing and RNP-depleted matrices analyzed by sequential fractionation and resinless section electron microscopy. J Cell Biol 1986; 102: 1654-1665.
100. Smith HC, Ochs RL, Fernandez EA et al. Macromolecular domains containing nuclear protein p107 and U-snRNP protein p28: further evidence for the in situ nuclear matrix. Mol Cell Biochem 1986; 70:151-158.
101. Poznanovic G, Sevaljevic L. The retention of core ribonucleoproteins in the nuclear matrix isolated from nuclei treated with the phenantroline-copper complex. Cell Biol Int Rep 1986; 10:55-63.

102. Huang S, Deerinck TJ, Ellisman MH et al. In vivo analysis of the stability and transport of nuclear poly (A)+ RNA. J Cell Biol 1994; 126:877-899.
103. Jackson DA, Hassan AB, Errington RJ et al. Visualization of focal sites of transcription within human nuclei. EMBO J 1993; 12:1059-1065.
104. Balkzon RD, Brinkley BR. Tubulin interaction with kinetochore proteins: Analysis by in vitro assembly and chemical cross-linking. J Cell Biol 1987; 105:855-862.
105. Casiano CA, Landberg G, Ochs RL et al. Autoantibodies to a novel cell-cycle regulated protein that accumulates in the nuclear matrix during S phase and is localized in the kinetochores and spindle midzone during mitosis. J Cell Sci 1993; 106:1045-1056.
106. Liao H, Winkfein RJ, Mack J et al. CENP-F is a protein of the nuclear matrix that assembles onto kinetochores at late G2 and is rapidly degraded after mitosis. J Cell Biol 1995; 130:507-518.
107. Compton DA, Cleveland DW. NuMA, a nuclear protein involved in mitosis and nuclear reformation. Curr Opin Cell Biol 1994; 6:343-346.
108. Tawfic S, Ahmed K. Association of casein kinase 2 with nuclear matrix. Possible role in nuclear matrix protein phosphorilation. J Biol Chem 1994; 269:7489-7493.
109. Eisenman RN, Hann SR. Proteins expressed by the c-myc oncogene in lymphomas of human and avian origin. Proc R Soc Lond Biol Sci 1985; 226:73-78.
110. Eisenman RN, Tachibana CY, Abrams HD et al. V-myc and c-myc-encoded proteins are associated with the nuclear matrix. Mol Cell Biol 1985; 5:14-126.
111. Van Straaten JP, Rabbits TH. The c-myc protein is associated with the nuclear matrix through specific metal interaction. Oncogene Res 1987; 1:221-228.
112. De Greve J, Battey J, Fedorko J et al. The human L-myc gene encodes multiple nuclear phosphoproteins from alternatively processed mRNAs. Mol Cell Biol 1988; 8:4381-4388.
113. Waitz W, Loidl P. Cell cycle dependent association of c-myc protein with the nuclear matrix. Oncogene 1991; 6:29-35.
114. Boyle WJ, Lampert MA, Li AC et al. Nuclear compartmentalization of the v-myb oncogene product. Mol Cell Biol 1985; 5:3017-3023.
115. Klempnauer KH, Sippel AE. Subnuclear localization of proteins encoded by the oncogene v-myb and its cellular homologue c-myb. Moll Cell Biol 1986; 6:62-69.
116. Klempnauer KH. Interaction of myb proteins with nuclear matrix in vitro. Oncogene 1988; 2:545-551.
117. Lipsick JS, Baluda MA. The myb oncogene. Gene Amplif Anal 1986; 4:73-98.

118. Slamon D, Boone TC, Murdock DC. Studies of the human c-myb gene and its product in human acute leukemias. Science 1986; 233:347-351.
119. Greenfield I, Nickerson J, Penman S et al. Human papillomavirus 16 E7 protein is associated with the nuclear matrix. Proc Natl Acad Sci USA 1991; 88:11217-11221.
120. Tommasino M, Contorni M, Cavalieri F. HPV16 E7 phosphorylation in fission yeast: characterization and biological effects. Gene 1992; 111:93-98.
121. Puvion E, Duthu A, Harper F et al. Intranuclear distribution of SV40 large T-antigen and transformation-related protein p53 in abortively infected cells. Exp Cell Res 1988; 177:73-89.
122. Schmitt RC, Fahnestock ML, Lewis JB. Differential nuclear localization of the major adenovirus type 2 E1a proteins. J Virol 1987; 61:247-255.
123. Hendzel MJ, Davie JR. Nuclear distribution of histone deacetylase: A marker enzyme for the internal nuclear matrix. Biochim Biophys Acta 1992; 1130:307-313.
124. Hendzel MJ, Sun J-M, Chen HY et al. Histone acetyltransferase is associated with the nuclear matrix. J Biol Chem 1994; 269: 22894-22901.
125. Davie JR, Hendzel MJ. Multiple functions of Dynamic histone acetylation. J Cell Biochem 1994; 55:98-105.
126. Portoles M, Faura M, Renau-Piqueras J et al. Nuclear calmodulin/62kDa calmodulin-binding protein complexes in interphasic and mitotic cells. J Cell Sci 1994; 107:3601-3614.
127. Barrack ER, Coffey DS. The specific binding of estrogens and androgens to the nuclear matrix of sex hormone responsive tissues. J Biol Chem 1980; 255:7265-7275.
128. Metzger DA, Korach KS. Cell-free interaction of the estrogen receptor with mouse uterine nuclear matrix: Evidence of saturability, specificity, and resistance to KCl extraction. Endocrinology 1990; 126:2190-2195.
129. Barrack ER. The nuclear matrix of the prostate contains acceptor sites for androgen receptors. Endocrinology 1983; 113:430-432.
130. Colvard DS, Wilson EM. Androgen receptor binding to nuclear matrix in vitro and its inhibition by 8S androgen receptor binding factor. Biochemistry 1984; 23:3479-3486.
131. Van Steensen B, van Haarst AD, deKloet ER et al. Binding of corticosteroid receptors to rat hippocampus nuclear matrix. FEBS Lett 1991; 292:229-231.
132. Van Steensen B, Jenster G, Damm K et al. Domains of the human androgen receptor and glucocorticoid receptor involved in binding to the nuclear matrix. J Cell Biochem 1995; 57:465-478.
133. Hora J, Horton MJ, Toft DO et al. Nuclease resistance and the enrichment of native nuclear acceptor sites for the avian oviduct progesteron receptor. Proc Natl Acad Sci USA 1986; 83:8839-8843.

134. Schuchard M, Landers JP, Punkay-Sandhu N et al. Nuclear matrix localization and specific matrix DNA binding factor-1 of the avian oviduct progesteron receptor. Biochemistry 1991; 30:9516-9522.
135. Barrack ER, Coffey DS. The role of the nuclear matrix in steroid hormone action. Biochem Actions Horm 1983; 10:23-90.
136. Barrack ER. Localization of steroid hormone receptors in the nuclear matrix. In: Clark CR, ed. Steroid Hormone Receptors: Their Intracellular Localization. Chichester, England: Horwood, 1987; 86-127.
137. Barrack ER. Specific association of androgen receptors and estrogen receptors with the nuclear matrix: summary and perspectives. In: Moudgil VK, ed. Recent Advances in Steroid Hormone Action. New York: de Gruyter, 1987:85-107.
138. Replogle-Schwab TS, Pienta KJ, Getzenberg RH. The utilization of nuclear matrix proteins for cancer diagnosis. Crit Rew Eukar Gene Expr 1996; 6:103-113.
139. Keesee SK, Briggman JV, Thill G et al. Utilization of nuclear matrix proteins for cancer diagnosis. Crit Rew Eukar Gene Expr 1996; 6:189-214.
140. Cook PR, Brazell IA, Jost E. Characterization of nuclear structures containing superhelical DNA. J Cell Sci 1976; 22:303-324.
141. Staufenbiel M, Deppert W. Preparation of nuclear matrices from cultured cells: Subfractionation of nuclei in situ. J Cell Biol 1984; 98:1886-1894.
142. Jackson DA, Cook PR. Transcription occurs at a nucleoskeleton. EMBO J 1985; 4:919-925.
143. Jackson DA, Cook PR. Replication occurs at a nucleoskeleton. EMBO J 1986; 5:1403-1410.
144. Mirkovich J, Mirault E, Laemmli UK Organization of the higher order chromatin loop: specific DNA attachment sites on nuclear scaffold. Cell 1984; 39:323-332

CHAPTER 2

What Happens with the Nuclear Matrix During Mitosis

2.1. GENERAL COMMENTS

The simplest answer to the question posed in the title of this chapter is that the nuclear matrix disintegrates during mitosis. This, indeed, happens with some parts of the nuclear matrix. An obvious example is the nuclear lamina which disassembles at the beginning of mitosis and reassembles after its completion.[1,2] However, upon closer examination the fate of distinct parts of the nuclear matrix during mitosis can be found to be very different. Some structural components of the interphase nuclear matrix seem to persist through mitosis without considerable changes. These are, for example, replication foci. Being pulse-labeled with bromodeoxiuridine during S-phase, the replication foci can be immunostained on metaphase chromosomes and, furthermore, in the daughter interphase nuclei after completion of mitosis.[3] Clusters of interchromatin granules and nuclear bodies also seem to retain their integrity in mitotic cells.[4,5] The nuclear matrix as a whole is, however, disassembled.

Several specific areas (or compartments) can be recognized in mitotic cells (Fig. 2.1). These are chromosomes, perichromosomal layer, mitotic spindle and cytoplasmic compartment. Recent evidence indicates that during mitosis the nuclear matrix proteins are nonrandomly distributed among the above mentioned compartments, as described below.

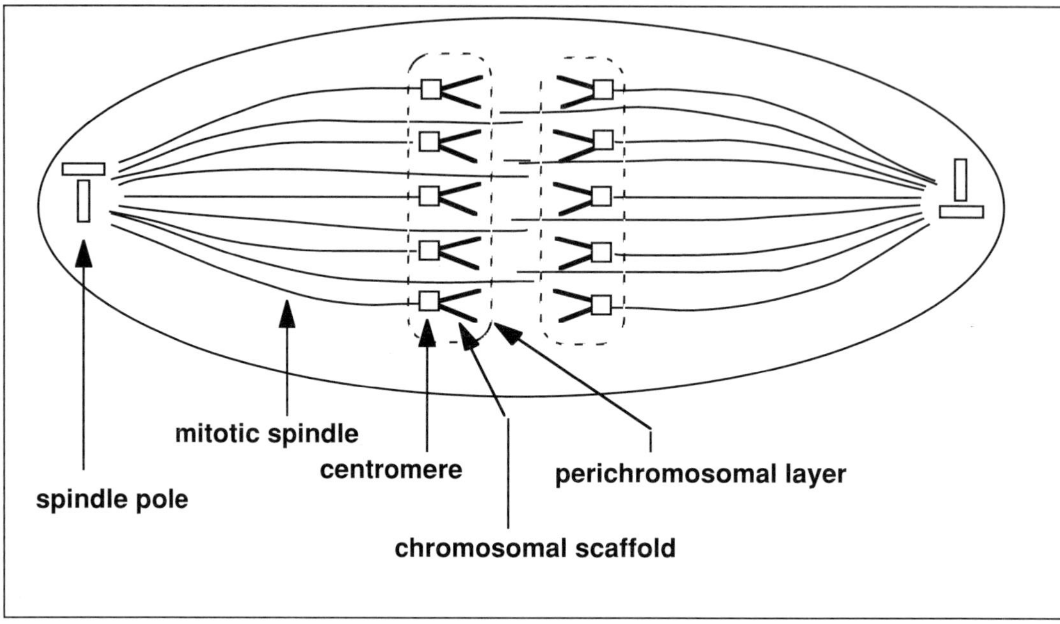

Fig. 2.1. A scheme illustrating compartmentalization of a mitotic cell.

2.2. CHROMOSOMES AND PERICHROMOSOMAL LAYER

A part of the nuclear matrix is likely to undergo a certain structural reorganization giving rise to the protein scaffolds of metaphase chromosomes[6] (Fig. 2.2). The scaffolds may be isolated by an experimental procedure similar to this used for the isolation of nuclear matrices (i.e., high-salt extraction of nuclease-treated metaphase chromosomes followed by separation of fast sedimenting structures using differential centrifugation). Keeping in mind the fact that scaffolds of metaphase chromosomes are resistant to extraction with 2M NaCl solution, they can also be referred to as chromosomal matrices. Their protein composition constitutes a matter of discussion. Initial studies revealed a relatively complex composition of chromosomal scaffolds.[6-10] However, it is difficult to rule out the possibility that scaffolds characterized in the above studies were heavily contaminated with cytoplasmic components and components of the perichromosomal layer. Lewis and Laemmli[11] established the presence in chromosomal scaffolds of two major polypeptide components with molecular weights of 170 kDa and 135 kDa termed SC I and SC II, respectively.[11] SC I (170 kDa) was identified with DNA topoisomerase II[12-14] which is also a major component of the interphase nuclear matrix. It should be mentioned that

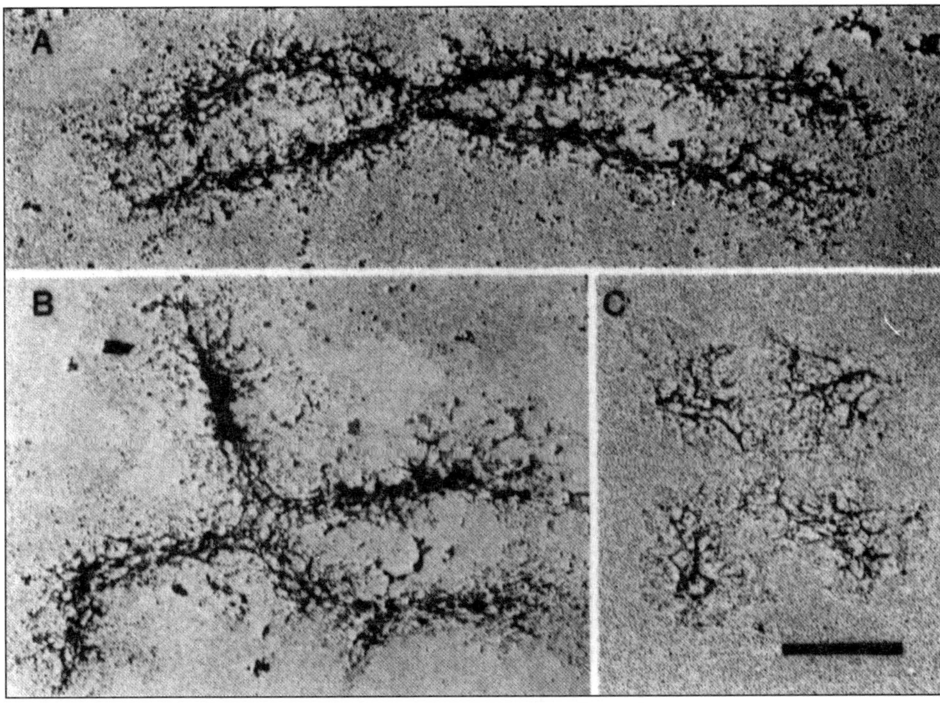

Fig. 2.2. Electron micrographs of the chromosomal scaffold. (A and B) Chromosomes were treated with micrococcal nuclease and then diluted into 2M NaCl, incubated at 4°C and spread with cytochrome c. (C) An example of a scaffold purified on a sucrose gradient. Bar represents 2 μm. (Reprinted with permission of K.W. Adolph from Adolph KW, Cheng SM, Paulson JR et al. Proc Natl Acad Sci USA 1977; 74:4937-4941. Copyright to individual articles retained by the authors).

structural role of the DNA topoisomerase II in a chromosomal scaffold has been questioned by some recent observations,[15,16] although the necessity of this enzyme for the condensation of metaphase chromosomes is well proven by experiments on chromosome assembly in vitro.[15,17] The gene encoding the second of the chromosomal matrix/scaffold major proteins (SC II) has recently been cloned and characterized.[18] The SC II protein has turned out to be a member of the SMS family of coiled-coil proteins. These proteins are not represented in the interphase nuclear matrix, although they play an important role in maintaining the integrity of the metaphase chromosome scaffold (for a review see refs. 19, 20). It is interesting that SC II can form a stable complex with topoisomerase II and with another presently uncharacterized 48 kDa protein.[21] This complex binds DNA with a more stringent sequence specificity than does topoisomerase II.[21]

In contrast to the nuclear matrix, the scaffold of a metaphase chromosome may be visualized even in nonextracted chromosomes by silver staining[22] or by immunostaining with antibodies against topoisomerase II.[13,14] Recent studies demonstrate that in nonextracted chromosomes the scaffolds are helically coiled and that scaffolds of sister chromatids have opposite helical handedness.[23] Under an electron microscope the chromosomal scaffold appears (depending on a preparation procedure) either as a fibrous network without prominent longitudinal (axial) fibers or as a denser rod-like structure.[6,24] Under certain experimental conditions it is possible to observe the fragmentation of scaffolds into several dense parts.[25] These parts of the scaffold remain connected to each other by DNA. Digestion of DNA by nucleolytic enzymes results in complete disintegration of chromosomes. Although these observations were interpreted in terms of artificial creation of scaffolds[25] in the course of isolation of metaphase chromosomes, in truth, they only show that the scaffold is not a uniform structure and that some of its regions are more labile than the others. Recent observations of Saitoh and Laemmli suggest that the folding path of the scaffold within a metaphase chromosome may be different in R and G/Q bands.[26] This may be one of the reasons for different levels of stability of the scaffold along the length of a metaphase chromosome.

The so-called perichromosomal layer constitutes a distinct compartment of mitotic cells. This layer encapsulates metaphase chromosomes insulating them from a direct contact with the cytoplasm. Recent evidence suggests that many components of the nuclear matrix are moved to this compartment during mitosis. Among them there are many nucleolar proteins,[27] including Ki-67 antigen,[28] perichromonucleolin,[29] fibrillarin[30] and B23 protein.[31] Desmin, cytokeratin-related p55 protein and hnRNP proteins are also transferred from the nuclear matrix to the perichromosomal compartment of mitotic cells.[32-36] In this compartment one can find certain proteins located in the periphery of interphase nuclei, such as lamin B,[1] perichromin[37] and P1 antigen[38] known also as peripherin.[39] The structural components of the nuclear skeleton (60-70 kDa nonlamina proteins) were also found in association with chromosomes during mitosis.[40] Similarly, a high molecular weight (240 kDa) nuclear matrix protein associated with 9 and 13 nm ribonucleoprotein filaments was found in the perichromosomal layer of mitotic cells.[41,42]

2.3. MITOTIC SPINDLE AND CYTOPLASMIC COMPARTMENT

Some matrix proteins play an important role in the organization of a mitotic spindle. The best known among them is NuMA protein which apparently plays a structural role in organization of interphase nuclear skeleton and contributes to the stabilization of mitotic spindle.[43-47] Recent data[48-50] indicate that there is a family of NuMA protein splice variants that includes centrophilin[51] and a 210 kDa protein known as Spindle-Pole-Nucleus (SPN) antigen.[52,53] Among other nuclear matrix proteins redistributed to the mitotic spindle we can mention here a Ca^{2+}/calmodulin-dependent protein kinase II which in many aspects behaves similarly to the NuMA protein,[54] and kinetochore proteins, CENP-B, CENP-C and CENP-F.[55-57] It should be stressed that with the exception of NuMA, all the above listed proteins seem to be simply deposited in the interphase nuclear matrix compartment without fulfilling any apparent function before the onset of mitosis. On the contrary, during mitosis these proteins play an essential role as inactivation of any of them by antibodies was found to cause a mitotic arrest. The SPN antigen is likely to act as a microtubular minus-end organizer.[52] Recently, another nuclear matrix protein has been cloned which apparently directs the assembly of the mitotic spindle. It has turned out to be a novel nuclear kinase related to the *raf* family of protooncogens.[58]

To finish the description of redistribution of nuclear matrix proteins in mitotic cells it is necessary to note that some nuclear matrix proteins are directed to the cytoplasmic compartment. In this compartment they seem to be transiently deposited on intermediate (namely vimentin) filaments.[59]

2.4. CONCLUSIONS

Although the nuclear matrix is disassembled in mitotic cells, the matrix proteins do not randomly diffuse within the whole cell volume. Most of them seem to be associated with chromosomes or to be deposited in the perichromosomal layer. Some nuclear matrix proteins fulfill specific functions in mitotic cells directing the organization of the mitotic spindle.

Related Review Articles

Nickerson JA, Penman S. Localization of nuclear matrix core filament proteins at interphase and mitosis. Cell Biol Int Rep 1992; 16:811-826.

Nigg EA. Assembly-disassembly of the nuclear lamina. Curr Opin Cell Biol 1992; 4:105-109.

Compton DA, Cleveland DW. NuMA, a nuclear matrix protein involved in mitosis and nuclear reformation. Curr Opin Cell Biol 1994; 6:343-346.

Hernandez-Verdun D, Gautier T. The chromosome periphery during mitosis. Bioessays 1994; 16:179-185.

He D, Zeng C, Brinkley BR. Nuclear matrix proteins as structural and functional components of mitotic apparatus. Int Rev Cytol 1995; 162B:1-74.

Cleveland DW. NuMA: a protein involved in nuclear structure, spindle assembly, and nuclear reformation. Trends Cell Biol 1995; 5:60-64.

References

1. Gerace L, Blobel G. The nuclear envelope lamina is reversibly depolymerized during mitosis. Cell 1980; 19:277-287.
2. Nigg EA. Assembly-disassembly of the nuclear lamina. Curr Opin Cell Biol 1992; 4:105-109.
3. Sparvoli E, Levi M, Rossi E. Replicon clusters may form structurally stable complexes of chromatin and chromosomes. J Cell Sci 1994; 107:3097-3103.
4. Turner BM, Franchi L. Identification of protein antigens associated with the nuclear matrix and with clusters of interchromatin granules in both interphase and mitotic cells. J Cell Sci 1987; 87:269-282.
5. Stuurman M, de Graaf A, Floore A et al. Monoclonal antibody recognizing nuclear matrix associated nuclear bodies. J Cell Sci 1992; 101:773-784.
6. Adolph KW, Cheng SM, Paulson JR et al. Isolation of a protein scaffold from mitotic HeLa cell chromosomes. Proc Natl Acad Sci USA 1977; 74:4937-4941.
7. Razin SV, Chernokhvostov VV, Roodin AV et al. Proteins tightly bound to DNA in the regions of DNA attachment to the skeletal structures of interphase nuclei and metaphase chromosomes. Cell 1981; 27:65-73.
8. Peters KF, Okada TA, Comings DE. Chinese hamster nuclear proteins. An electrophoretic analysis of interphase, metaphase and nuclear matrix preparations. Eur J Biochem 1982; 129:221-232.
9. Gooderham K, Jeppesen P. Chinese hamster metaphase chromosomes isolated under physiological conditions. A partial characterization of associated non-histone proteins and protein cores. Exp Cell Res 1983; 144:1-14.

10. Peick AC, van der Velden HM, Rijken AA et al. Protein composition of the chromosomal scaffold and interphase nuclear matrix. Chromosoma 1985; 91:137-144.
11. Lewis C, Laemmli UK. Higher order metaphase chromosome structure: evidence for metalloprotein interactions. Cell 1982; 29: 171-181.
12. Earnshaw WC, Halligan B, Cooke CA et al. Topoisomerase II is a structural component of mitotic chromosome scaffolds. J Cell Biol 1985; 100:706-1715.
13. Earnshaw WC, Heck MMS. Localization of topoisomerase II in mitotic chromosomes. J Cell Biol. 1985; 100:1716-1725.
14. Gasser SM, Laroche T, Falquet J et al. Metaphase chromosome structure: Involvement of topoisomerase II. J Mol Biol 1986; 188:613-629.
15. Hirano T, Mitchinson TJ. Topoisomerase-II does not play a scaffolding role in the organization of mitotic chromosomes assembled in *Xenopus* egg extracts. J Cell Biol 1993; 120:601-612.
16. Swedlow JR, Sedat JW, Agard DA. Multiple chromosomal populations of topoisomerase II detected in vivo by time-lapse, three dimensional wide-field microscopy. Cell 1993; 73:97-108.
17. Adachi Y, Luke M, Laemmli UK Chromosome assembly in vitro: topoisomerase II is required for condensation. Cell 1991; 64:137-148.
18. Saitoh N, Goldberg IG, Wood ER et al. ScII: an abundant chromosome scaffold protein is a member of a family of putative ATPases with an unusual predicted tertiary structure. J Cell Biol 1994; 79:449-458.
19. Hirano T. Biochemical and genetic dissection of mitotic chromosome condensation. Trends Biochem Sci 1995; 20:357-361.
20. Gasser SM. Coiling up chromosomes. Current Biol 1995; 5:357-360.
21. Ma X, Saitoh N, Curtis PJ. Purification and characterization of a nuclear DNA-binding factor complex containing topoisomerase II and chromosome scaffold protein 2. J Biol Chem 1993; 268: 6182-6188.
22. Howell WM, Hsu TC. Chromosome core structure revealed by silver staining. Chromosoma 1979; 73:61-66.
23. Boy de la Tour E, Laemmly UK. The metaphase scaffold is helically folded: Sister chromatids have predominantly opposite helical handedness. Cell 1988; 55:937-944.
24. Paulson JR. Scaffold morphology in histone-depleted HeLa metaphase chromosomes. Chromosoma 1989; 97:289-295.
25. Hadlaczky G, Sumner AT, Ross A. Protein-depleted chromosomes. II Experiments concerning the reality of chromosome scaffolds. Chromosoma 1981; 81:557-567.

26. Saitoh H, Laemmli UK. Metaphase chromosome structure: Bands arise from a differential folding path of the high AT-rich scaffold. Cell 1994; 76:609-622.
27. Gautier T, Dauphin-Villemant C, Andre C et al. Identification and characterization of a new set of nucleolar ribonucleoproteins which line the chromosomes during mitosis. Exp Cell Res 1992; 200:5-15.
28. Verheijen R, Kuijpers HJ, van Driel R et al. Ki-67 detects a nuclear matrix-associated proliferation-related antigen. II. Localization in mitotic cells and association with chromosomes. J Cell Sci 1989; 92:531-540.
29. Shi L, Zumei N, Shi Z et al. Involvement of a nucleolar component, perichromonucleolin, in the condensation and deconstruction of chromosomes. Proc Natl Acad Sci USA 1987; 84:7953-7956.
30. Yasuda Y, Maul GG. A nucleolar auto-antigen is part of a major chromosomal surface component. Chromosoma 1990; 99:152-160.
31. Tawfic S, Goueli SA, Olson MO et al. Androgenic regulation of the expression and phosphorilation of prostatic nucleolar protein B23. Cell Mol Biol Res 1993; 39:43-51.
32. Chai LS, Sandberg AA. Chromosomes and their relationship to nuclear components during the cell cycle in Chinese hamster cells. Cell Tissue Res 1988; 251:197-204.
33. Kamei H. Location of nuclear antigen(s) recognized by DSB389 MAb, a monoclonal antibody against desmin, observed by confocal laser scanning fluorescence microscopy. Cell Biol Int Rep 1990; 14:727-736.
34. Alique R, Bastos R, Serratosa J et al. Increase in a 55-kDa keratin-like protein in the nuclear matrix of rat liver cells during proliferative activation. Exp Cell Res 1990; 186:346-353.
35. He DC, Martin T, Penman S. Localization of heterogeneous nuclear ribonucleoprotein in the interphase nuclear matrix core filaments and on perichromosomal filaments in mitosis. Proc Natl Acad Sci USA 1991; 88:7469-7473.
36. Hernandez-Verdun D, Gautier T. The chromosome periphery during mitosis. Bioessays 1994; 16:179-185.
37. McKeon FD, Tuffanelli DL, Kobayashi S et al. The redistribution of a conserved nuclear envelope protein during the cell cycle suggests a pathway for chromosome condensation. Cell 1984; 36:83-92.
38. Chaly N, Bladon T, Setterfield G et al. Changes in distribution of nuclear matrix antigens during the mitotic cell cycle. J Cell Biol 1984; 99:661-671.
39. Chaly N, Brown DL. The prometaphase configuration and chromosome order during mitosis. J Cell Sci 1988; 91:325-335.
40. Noaillac-Depeyre J, Azum MC, Geraud ML et al. Distribution of nuclear matrix proteins in interphase CHO cells and rearrangements during the cell cycle. An ultrastructural study. Biol Cell 1987; 61:23-32.

41. Nickerson JA, Penman S. Localization of nuclear matrix core filament proteins at interphase and mitosis. Cell Biol Int Rep 1992; 16:811-826.
42. Nickerson JA, Krockmalnic G, Wan KM et al. A normally masked nuclear matrix antigen that appears at mitosis on cytoskeleton filaments adjoining chromosomes, centrioles and midbodies. J Cell Biol 1992; 116:977-987.
43. Compton DA, Cleveland DW. NuMA is required for the proper completion of mitosis. J Cell Biol 1993; 120:947-957.
44. Yang CH, Snyder M. The nuclear-mitotic apparatus protein (NuMA) is important in the establishment and maintenance of the bipolar mitotic spindle apparatus. Mol Biol Cell 1992; 3:1259-1267.
45. Compton DA, Cleveland DW. NuMA, a nuclear protein involved in mitosis and nuclear reformation. Curr Opin Cell Biol 1994; 6:343-346.
46. Zeng C, He D, Brinkley BR. Localization of NuMA protein isoforms in the nuclear matrix of mammalian cells. Cell Motil Cytoskeleton 1994; 29:167-176.
47. Zeng C, He D, Berget SM et al. Nuclear-mitotic apparatus protein: a structural interphase between the nucleoskeleton and RNA splicing. Proc Natl Acad Sci USA 1994; 91:1505-1509.
48. Tang TK, Tang CC, Chen Y et al. Nuclear proteins of the bovine esophageal epithelium II: the NuMA gene gives rise to multiple mRNAs and gene products reactive with monoclonal antibody W1. J Cell Sci 1993; 104:249-260.
49. Compton DA, Szilak I, Cleveland DW Primary structure of NuMA, an intranuclear protein that defines a novel pathway for segregation of proteins at mitosis. J Cell Biol 1992; 116:1395-1408.
50. Yang CH, Lambie EJ, Synder M. NuMA: an unusually long coiled coil related protein in the mammalian nucleus. J Cell Biol 1992; 116:1303-1317.
51. Tousson A, Zeng C, Brinkley BR et al. Centrophilin: a novel mitotic spindle protein involved in microtubule nucleation. J Cell Biol 1991; 112:427-440.
52. Kallajoki M, Weber K, Osborn M. Ability to organize microtubules in taxol-treated mitotic PtK2 cells goes with SPN antigen and not with the centrosome. J Cell Sci 1992; 102:91-102.
53. Kallajoki M, Osborn M. Gel electrophoresis of nuclear matrix fractions isolated from different human cell lines. Electrophoresis 1994; 15:520-528.
54. Ohta Y, Ohta T, Miyamota E. Ca^{2+}/calmodulin-dependent protein kinase II: localization in the interphase nucleus and the mitotic apparatus of mammalian cells. Proc Natl Acad Sci USA 1990; 87:5341-5345.
55. Liao H, Winkfein RJ, Mack J et al. CENP-F is a protein of the nuclear matrix that assembles onto kinetochores at late G2 and is rapidly degraded after mitosis. J Cell Biol 1995; 130:507-518.

56. Balkzon RD, Brinkley BR. Tubulin interaction with kinetochore proteins: Analysis by in vitro assembly and chemical cross-linking. J Cell Biol 1987; 105, 855-862.
57. Casiano CA, Landberg G, Ochs RL et al. Autoantibodies to a novel cell-cycle regulated protein that accumulates in the nuclear matrix during S phase and is localized in the kinetochores and spindle midzone during mitosis. J Cell Sci 1993; 106:1045-1056.
58. Johansen KM, Johansen J. Dynamic redistribution during mitosis of a novel nuclear matrix protein with raf-kinase homology. J Cell Biochem 1995; Suppl. 21B:139.
59. Marugg RA. Transient storage of a nuclear matrix protein along intermediate-type filaments during mitosis: a novel function of cytoplasmic intermediate filaments. J Struct Biol 1992; 108:129-139.

CHAPTER 3

Periodical Attachment of Chromosomal DNA to the Nuclear Matrix Organizes this DNA in Large Closed Loops

3.1. IDENTIFICATION OF CHROMOSOMAL DNA LOOPS BOUND TO THE NUCLEAR MATRIX

In early experiments rat livers were often used for isolation of nuclear matrices.[1-3] Liver cells contain active nucleases which extensively digested DNA already at the step of isolation of nuclei. Furthermore, isolated nuclei were either stored overnight at 0°C or incubated for a short time at 37°C in order to permit partial autodigestion of DNA before any extractions were carried out. That is why no digestion with an exogenous nuclease was necessary for subsequent solubilization of a major portion of chromatin by a low salt and a high salt extractions. A different situation was, however, observed when attempts were made to prepare nuclear matrices from other cells. It turned out that a significant, if not the major, portion of DNA could not be separated from the residual nuclear structures even by 2M NaCl extraction unless the nuclei were pretreated with exogenous nucleases (see, for example, ref. 4).

The most interesting observations were, perhaps, made by Cook and collaborators who subjected whole cells to extraction with

2M NaCl solution containing also EDTA and a nonionic detergent.[5-8] Upon this extraction, the nuclear DNA was not simply released into the solution but remained organized into loops bound to the high-salt-insoluble nuclear remnants which were called "nuclear ghost" or "nuclear cage" and which presently may be identified with the "nuclear matrix." The whole DNA-containing complex was called "nucleoid". DNA loops in nucleoids possess the properties of closed domains, i.e., from the topological point of view they are analogous to covalently closed circular DNA. Disruption of nucleosomes by 2M NaCl extraction gives rise to negative supercoiling of the loop DNA. Supercoiled DNA loops can be directly observed with an electron microscope in spread nucleoid preparations.[9,10] The density of negative supercoils in the loop DNA can be decreased by treating nucleoids with ethidium bromide which intercalates into the DNA and causes partial unwinding of the DNA double helix.[6,7,11] By titration with an increasing concentration of ethidium bromide it is possible to eliminate all negative supercoils and, furthermore, to induce positive supercoiling in closed DNA loops. Changes in the superhelical density influence the sedimentation properties of nucleoids. As a result, nucleoids sediment biphasically in gradients containing intercalators. When DNA loops are completely relaxed, the sedimentation coefficient is minimal because in this state they have the least compact configuration (Fig. 3.1).

Whatever the interactions which hold the ends of DNA loops may be, these interactions are more stable than the interactions between most nuclear matrix proteins. Indeed, a folded (i.e., organized into loops) genome can be released from cells treated with an ionic detergent sodium dodecyl sulphate. This treatment causes disintegration of a morphologically distinct nuclear matrix, but it does not injure the topological locks at the loop ends.[12,13] The integrity of metaphase chromosome scaffolds may also be affected by some treatments which do not influence the mode of DNA organization into loops.[14]

The results of analysis of DNA packaging in nucleoids are not the only ground to suppose that eukaryotic DNA is arranged into loops fixed at some proteinaceous structure. Recent data of Nicolini and collaborators suggest that DNA is organized into closed domains also in living cells.[15] As has been mentioned above, large DNA loops were directly visualized by means of electron microscopy

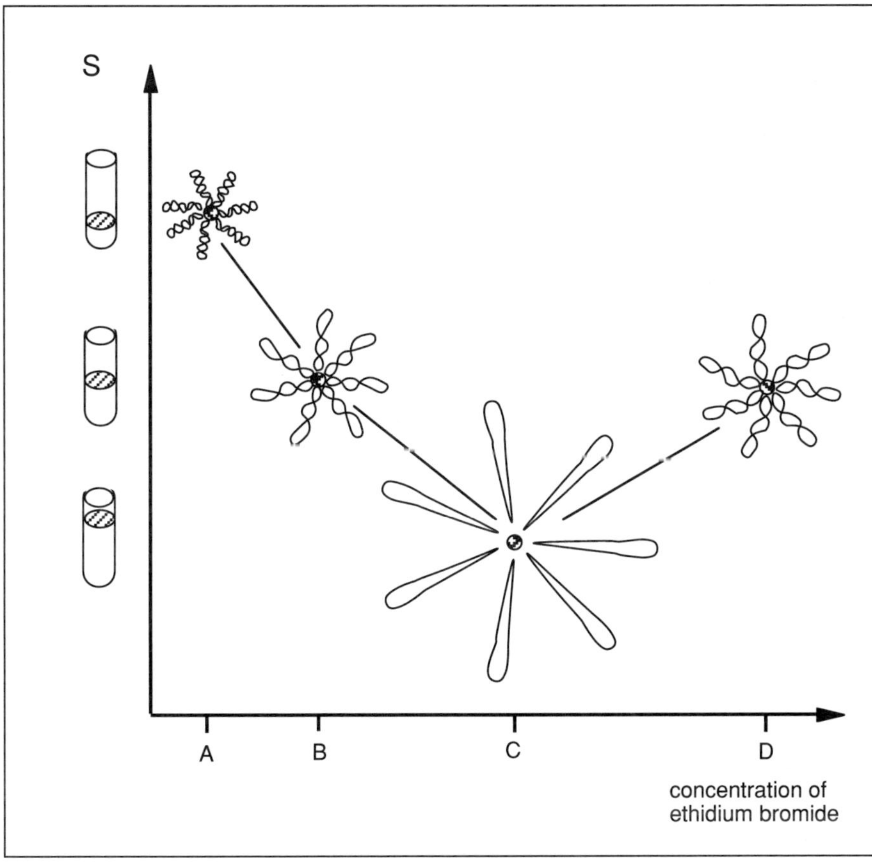

Fig. 3.1. A scheme illustrating the biphasic dependence of sedimentation coefficient of nucleoids on concentration of the ethidium bromide in the medium. With an increase of the ethidium bromide concentration, the negatively supercoiled DNA loops are gradually unwound (A-C) until a full loss of supercoils (C). Thereafter positive supercoils are introduced (D). Unwinding of DNA loops leads to an increase of the diameter of the loop halo (A-C). Correspondingly, the sedimentation coefficient of nucleoids decreases due to the "parachute" effect. Positive supercoiling of DNA loops introduced at high concentrations of ethidium bromide decreases the diameter of nucleoid and increases its sedimentation coefficient.

in spread metaphase chromosomes and interphase nuclei pre-extracted with 2M NaCl solution[9,10,16-18] (Fig. 3.2). If DNA was not nicked when preparing spreads, it was possible to see that the loops were supercoiled.[9,10,16] Similar-sized DNA loops (taking into account the level of compactization achieved by DNA packaging into nucleosomes) were observed in metaphase chromosomes not extracted with a concentrated salt solution.[19,20] Analysis of the size distribution of DNA fragments remaining bound to the nuclear

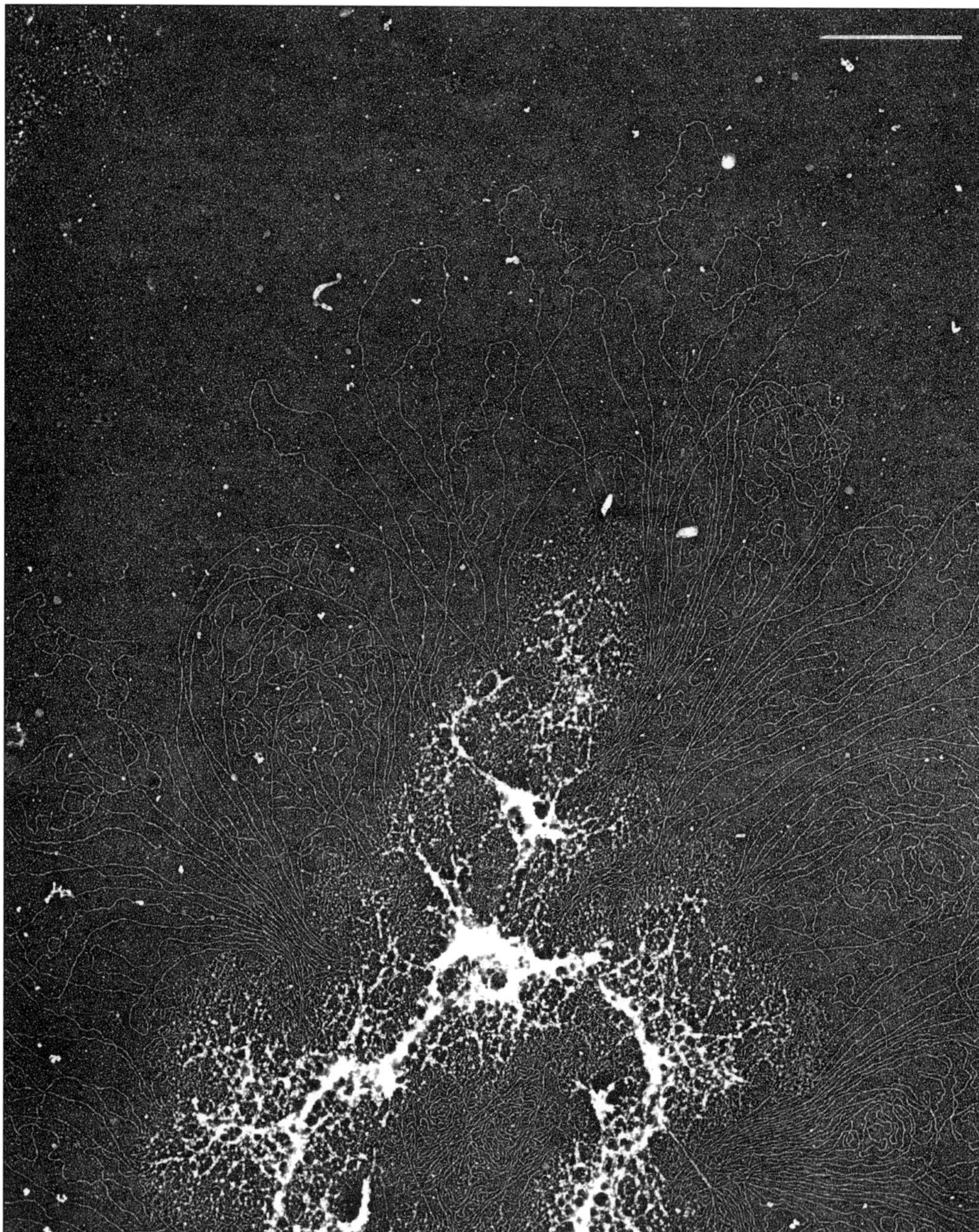

Fig. 3.2. DNA loops bound to the nuclear matrix. The white bar corresponds to 1 μm. The electron microscopic photo shown in the figure was kindly given by R. Hancock especially for publication in this book.

matrix and released into solution upon high-salt extraction of nuclei treated for increasing time intervals with nucleases also indicated that DNA was periodically attached to some internal nuclear structure.[4,21] Indeed, the size of released DNA fragments was limited in agreement with the prediction that it cannot exceed the size of a loop fixed at the nuclear matrix. The average sizes of nuclear matrix-bound DNA fragments correlated linearly with the percentage of DNA recovered from the nuclear matrix.

Cytochemical studies demonstrated that the majority of DNA loop anchorage sites resided in the internal matrix, although attachments to the nuclear lamina were also observed.[22] As has been discussed in chapter 1, the internal part of the nuclear matrix is relatively unstable and may be lost partially or even completely in the course of chromatin extraction. This may be one of the reasons for different estimates of the average size of DNA loops made in experiments with the removal of chromatin by different extraction procedures.[23]

3.2. ESTIMATION OF AN AVERAGE LOOP SIZE

Several approaches have been used to estimate an average size of chromosomal DNA loops. These approaches can be divided into three groups:

3.2.1. Nucleoid relaxation analysis

As has been discussed before, the sedimentation properties reflecting the apparent size of nucleoids depend on the superhelical density of DNA loops in nucleoids. When DNA loops are completely relaxed, the sedimentation coefficient is minimal because this is the least compact configuration of a nucleoid. Complete relaxation of DNA loops can be achieved not only by the incorporation of an appropriate amount of an intercalating agent into DNA, but also by introducing a single-stranded DNA scission (nick) in each loop. Once this is done, it is necessary simply to determine an average distance between the nicks. It is easy to see that in the simplest case, i.e., when entire DNA is arranged into loops of relatively the same size, this distance must be equal to the double size of an average loop.[11,24,25] If, however, the sizes of DNA loops vary significantly, or if there are several populations of loops differing in their average sizes, the sedimentation properties of nucleoids will be determined by relaxation of the largest loops. Hence, an average loop size may be easily overestimated with the above described approach. A correct estimation of the average distance between the nicks introduced with a frequency of approximately one nick per DNA loop (i.e., estimation of the average size of 50-200 kb single-stranded DNA fragments) also constitutes a certain experimental problem. In some initial studies, this distance was calculated indirectly based on the dose of X-ray irradiation used for introducing nicks in loop DNA. This indirect estimation could cause serious mistakes.

There is another and simpler version of estimation of the DNA loop size by the nucleoid relaxation analysis (see refs. 26, 27 for example). Cells grown on coverslips are extracted with 2M NaCl solution containing also a nonionic detergent. After this treatment the nuclei remain attached to the coverslips through cytoskeletal elements and histone-depleted DNA forms a halo of supercoiled loops attached to the nuclear matrix. After relaxation of the loops by introducing single-stranded DNA scissions or by treatment with topoisomerase I, the diameter of the halo (which can be measured under a fluorescence microscope) will be equal to an average loop size.

An average size of DNA loops estimated by the above-described variants of the nucleoid relaxation analysis turned out to be within the range from 20-200 kb depending on the type of cells.[11,24-27] (for a review see refs. 28, 29). Again, it is necessary to keep in mind that the largest loops determine the halo diameter. Furthermore, the portion of DNA nonorganized into loops (if any) cannot be accurately estimated by any variant of the nucleoid relaxation approach. According to the data of Vogelstein and collaborators,[27] this portion (DNA residing within the nuclear matrix) may constitute as much as 20% of the whole nucleoid DNA.

3.2.2. Calculation of an average loop size on the basis of a correlation between the percentage of DNA in the nuclear matrix and the average size of nuclear matrix DNA fragments

Treatment of nucleoids with nucleases producing double stranded DNA breaks results in a gradual release of DNA from the nuclear matrix. Within an area of 2-50 kb, the average size of DNA fragments remaining bound to the nuclear matrix after limited DNA cleavage correlates linearly with the percentage of DNA recovered in the nuclear matrix.[4,23] On the basis of this correlation it is possible to estimate an average size of DNA loops by approximation to the starting point of digestion. Clearly, the interpretation of the above correlation is based on the assumption that all DNA is organized in more or less uniform loops and that the attachment regions are either small enough or are protected from nucleases. At least the second assumption may be incorrect because no linear correlation between the percentage of DNA in nuclear matrices and the average size of nuclear matrix DNA fragments was observed when less than 2% of nuclear DNA remained in the nuclear matrices (Razin et al, unpublished results).

An average size of DNA loops estimated on the basis of a correlation between the percentage of DNA in the nuclear matrix and the average size of nuclear matrix DNA fragments was found to be within the range of 40-80 kb.[4,30-32] This approach can also be used to estimate an average size of DNA loops in nuclei which have not been extracted with a high salt solution. In this case the cleaved-off portions of loop DNA (still arranged into nucleosomes) can be electroeluted from nuclei in a "physiological buffer."[23] The average size of DNA loops in HeLa cell nuclei not extracted with 2M NaCl was found to be 86 kb,[23] i.e., it did not differ significantly from the average sizes of DNA loops in high-salt-extracted nucleoids.

3.2.3. Measurement of the loop length on electron microscopic pictures of spread nucleoids and histone-depleted metaphase chromosomes

Visualization of DNA loops in histone-depleted metaphase chromosomes[13] and interphase nuclei[14] (Fig. 3.2) provides the possibility to measure their length directly. In HeLa cells, loops ranging in size from 30-90 kb were detected.[17] One should, however, keep in mind that during spreading the DNA may be mechanically broken and thus the probability to lose longer loops will be relatively higher. Hence, an average loop size determined by direct measurements of the DNA loop length on electron microscopic pictures may be underestimated. Again, the portion of DNA residing within the scaffolding elements (if any) cannot be estimated.

3.3. TWO LEVELS OF DNA PACKAGING IN NUCLEOIDS

It has been reported by several authors that the usage of metal chelators or thiol reagents in the course of isolation of histone-depleted nuclei or metaphase chromosomes results in a sharp increase of the diameter of the DNA loop halo.[33-35] Correspondingly, the sedimentation coefficient of nucleoids decreased from 18000 S to 8500 S. The effect was completely reversible and compact nucleoids could be obtained upon addition of Cu^{2+} ions up to a concentration of 10^{-6} M. Other divalent ions, such as Mn^{2+}, Co^{2+}, Zn^{2+} and Hg^{2+}, were unable to substitute Cu^{2+}. The compact configuration, however, could be reestablished (at least partially) as a result of incubation in a Ca^{2+}-containing buffer at 37°C.[35] These results were interpreted in terms of the existence of two different groups of DNA loop anchorage sites on the nuclear matrix, one of these groups being organized with the participation of Cu^{2+} ions.[33,34] Some observations[36] suggest that Cu^{2+}-dependent and Cu^{2+}-independent associations of DNA with the nuclear matrix may involve different kinds of DNA fragments. In mouse Erlich ascite carcinoma cell nuclei, Cu^{2+}-dependent associations with the nuclear matrix were found to protect from nuclease digestion DNA fragments ranging in size from 50-150 bp. In contrast, Cu^{2+}-independent associations with the nuclear matrix protected from nuclease digestion long DNA fragments of about 10 kb. These fragments were composed of AT-rich satellite DNA sequences.[36] It was proposed that long protected fragments of nuclear matrix DNA originated from the sites of attachment of whole chromosomes to the nuclear lamina through telomeric and/or centromeric regions.[36]

3.4. ORGANIZATION OF CHROMOSOMAL DNA INTO LOOPS PERSISTS DURING MITOSIS

As mentioned in the beginning of this chapter, DNA loops can be directly visualized in spread metaphase chromosomes. It has been found in several independent studies that the average size of chromosomal DNA loops is the same in interphase nuclei and metaphase chromosomes.[4,29,37,38] The sequence complexity of nuclear matrix DNA isolated from interphase and metaphase cells is also very similar, if not the same.[4] Finally, recent experiments on mapping the DNA loop organization of nucleolar genes by the topoisomerase II-mediated DNA loop excision protocol (see chapter 5) have directly demonstrated that the mode of DNA organization into loops is preserved in metaphase chromosomes.[38]

3.5. CONCLUSIONS

Chromosomal DNA (or at least most of it) is periodically attached to a high-salt-insoluble nuclear matrix. As a result, the DNA is organized into large (20-200 kb, according to different estimations) closed loops. This organization persists during the whole cell cycle including mitosis. Whatever the nature of interactions of DNA with the nuclear matrix at loop anchorage sites is, these interactions are not affected by extraction with 2M NaCl solution. This extraction, however, disrupts nucleosomes generating an unconstrained superhelical tension within the loops.

Although several different approaches were used to determine an average size of DNA loops, the interpretation of experimental results was always based on the simplest assumption that the whole genome is organized into loops of more or less similar sizes. If a portion of the genome (even as large as 10 or 20%) were organized in a completely different fashion, this portion would be automatically excluded from the analysis by any of the above discussed experimental procedures. This is especially true of the procedures based on DNA relaxation in nucleoids that were used in most of the studies aimed to estimate the DNA loop size in different organisms. Hence, what was actually estimated in all these studies is not the average loop size, but rather the average size of the majority of uniformly organized loops.

The percentage of DNA nonorganized into loops cannot be estimated by any of the approaches used for determining an average loop size.

RELATED REVIEW ARTICLES

Hancock R. Topological organization of interphase DNA: the nuclear matrix and other skeletal structures. Biol Cell 1982; 46:105-122.

Jackson DA, Dickinson P, Cook PR. The size of chromatin loops in HeLa cells. EMBO J 1990; 9:567-571.

Jackson DA, Dolle A, Robertson G et al. The attachments of chromatin loops to the nucleoskeleton. Cell Biol Int Rep 1992; 16:687-696.

Razin SV, Gromova II, Iarovaia OV. Specificity and functional significance of DNA interaction with the nuclear matrix: New approaches to clarify the old questions. Int Rev Cytol 1995; 162B:405-448.

References

1. Berezney R, Coffey DS. Identification of a nuclear protein matrix. Biochem Biophys Res Commun 1974; 60:1410-1417.
2. Berezney R, Coffey DS. Nuclear protein matrix: association with newly synthesized DNA. Science 1975; 189:291-292.
3. Berezney R, Coffey DS. Nuclear matrix: Isolation and characterization of a framework structure from rat liver nuclei. J Cell Biol 1977; 73:616-637.
4. Razin SV, Mantieva VL, Georgiev GP. The similarity of DNA sequences remaining bound to scaffold upon nuclease treatment of interphase nuclei and metaphase chromosomes. Nucl Acids Res 1979; 17:1713-1735.
5. Cook PR, Brazell IA. Supercoils in human DNA. J Cell Sci 1975; 19:261-279.
6. Cook PR, Brazell IA. Conformational constrains in human DNA. J Cell Sci 1976; 22:287-302.
7. Cook PR, Brazell IA, Jost E. Characterization of nuclear structures containing superhelical DNA. J Cell Sci 1976; 22:303-324.
8. Cook PR, Brazell IA. Spectrofluorometric measurement of the binding of ethidium bromide to superhelical DNA from cell nuclei. Eur Biochem 1978; 84:464-477.
9. McCready SJ, Akrigg A, Cook PR. Electron microscopy of intact nuclear DNA from human cells. J Cell Sci 1979; 39:53-62.
10. Mullinger AM, Johnson RT. The organization of supercoiled DNA from human chromosomes. J Cell Sci 1979; 38:369-389.
11. Benyajati C, Worcel A. Isolation, characterization and structure of the folded interphase genome of *Drosophila* melanogaster. Cell 1976; 9:393-407.
12. Ide T, Nakane M, Anzai K et al. Supercoiled DNA folded by nonhistone proteins in cultured mammalian cells. Nature 1975; 258:445-447.
13. Nakane M, Ide T, Anzai K et al. Supercoiled DNA folded by nonhistone proteins in cultured mouse carcinoma cells. J Biochem 1978; 84:145-157.
14. Hadlaczky G, Sumner AT, Ross A. Protein-depleted chromosomes. II Experiments concerning the reality of chromosome scaffolds. Chromosoma 1981; 81:557-567.
15. Vergani L, Mascelti G, Nicolini C. Ethidium bromide intercalation and chromatin structure. Termichim Acta (in press).
16. Mullinger AM, Johnson RT. Packing DNA into chromosomes. J Cell Sci 1980; 46:61-86.
17. Paulson JR, Laemmli UK. The structure of histone-depleted metaphase chromosomes. Cell 1977; 12:815-828.
18. Hancock R, Hughes ME. Organization of DNA in the eukaryotic nucleus. Biol Cell 1982; 44:201-212.
19. Rattner JB, Hamkalo BA. Higher order structure in metaphase chromosomes. I. The 250 A fiber. Chromosoma 1978; 69:363-372.

20. Rattner JB, Hamkalo BA. Higher order structure in metaphase chromosomes. II. The relationship between the 250 A fiber, superbeads and beads-on-string. Chromosoma 1978; 69:373-379.
21. Igo-Kemennes T, Zachau HG. Domains in chromatin structure. Cold Spring Harb Symp Quant Biol 1977; 42:09-118.
22. Zini N, Mazzotti G, Santi P. Cytochemical localization of DNA loop attachment sites to the nuclear lamina and to the inner nuclear matrix. Histochemistry 1989; 91:199-204.
23. Jackson DA, Dickinson P, Cook PR. The size of chromatin loops in HeLa cells. EMBO J 1990; 9:567-571.
24. Hartwig M. Organization of mammalian chromosomal DNA: supercoiled and folded circular DNA subunits from interphase cell nuclei. Acta Biol Med Germ 1982; 37:421-432.
25. Nakane M, Ide T, Anzai K et al. Supercoiled DNA folded by nonhistone proteins in cultured mouse carcinoma cells. J Biochem 1978; 84:145-157.
26. Buongiorno-Nardelli M, Gioacchino M, Carri MT et al. A relationship between replicon size and supercoiled loop domains in the eukaryotic genome. Nature 1982; 298:100-102.
27. Vogelstein B, Small D, Robinson S et al. The nuclear matrix and the organization of nuclear DNA. In: Bekhor I, ed. Progress in Nonhistone Protein Research. Boca Raton, Florida: CRC Press, Inc., 1985; 2:115-129.
28. Hancock R. Topological organization of interphase DNA: the nuclear matrix and other skeletal structures. Biol Cell 1982; 46:105-122.
29. Hancock R, Boulikas T. Functional organization in the nucleus. Int Rev Cytol 1982; 79:165-214.
30. Berezney R, Buchholtz LA. Isolation and characterization of rat liver nuclear matrices containing high-molecular weight deoxyribonucleic acid. Biochemistry 1981; 20:4995-5002.
31. Mullenders LHF, Van Zeeland AA, Natarayan AT. Comparison of DNA loop size and supercoiled domain size in human cells. Mutat Res 1983; 112:245-252.
32. Ganguly A, Bagchi B, Bera M et al. Estimation of domain length of chicken erythrocyte chromatin. Biochem Biophys Res Commun 1983; 739:286-290.
33. Lebkowski JS, Laemmli UK. Evidence for two levels of DNA folding in histone-depleted HeLa interphase nuclei. J Mol Biol 1982; 156:309-324.
34. Lebkowski JS, Laemmli UK. Non-histone proteins and long-range organization of HeLa interphase DNA. J Mol Biol 1982; 156: 325-344.
35. Lewis CD, Laemmli UK. Higher order metaphase chromosome structure: evidence for metalloprotein interactions. Cell 1982; 29:171-181.

36. Razin SV, Chernokhvostov VV, Yarovaia OV. Organization of the sites for DNA attachment to the nonhistone proteinaceous nuclear skeleton. In: Bekhor I, ed. Progress in Nonhistone Protein Research. Boca Raton, FL: CRC Press, Inc., 1985; 2:91-114.
37. Warren AC, Cook PR. Supercoiling of DNA and nuclear conformation during the cell cycle. J Cell Sci 1978; 30:211-226.
38. Razin SV, Hancock R, Iarovaia O et al. Structural-functional organization of chromosomal DNA domains. Cold Spring Harbor Symp Quant Biol 1993; 58:25-35.

CHAPTER 4

FUNCTION-DEPENDENT AND PERMANENT INTERACTIONS OF DNA WITH THE NUCLEAR MATRIX

4.1. EXPERIMENTAL APPROACHES USED FOR ANALYSIS OF THE SPECIFICITY OF DISTRIBUTION OF UNIQUE DNA SEQUENCES WITHIN LOOPS

Since chromosomal DNA loops were first observed, the proposition has been discussed that they may correspond to the functional domains (or units) of the genome. In other words, it was proposed that loops might represent individual replicons or (and) transcriptions. With the aim to check this supposition, the specificity of DNA organization into loops was studied.

As a matter of fact, there are two different questions concerning the specificity of DNA organization into loops, namely: "Do the loop borders have specific positions in the genome?" and "What kind of sequences are located at the loop borders?" In this chapter we shall address only the first of the above questions while the second will be addressed in chapter 6. To find out whether the loop borders have specific positions in the genome, it is necessary first to isolate DNA fragments located close to the loop basements or even at the basements themselves. It is then possible to choose some genomic area and to study by hybridization experiments whether any part of this area is overrepresented in preparations of loop basements. Although seeming to be quite simple, the experiment completely depends on the availability of a correct procedure for isolation of loop basements. In a number of experiments, these basements were identified with nuclear matrix DNA, i.e., with DNA that can be isolated in complex with the nuclear matrix after mild or extensive digestion of nuclei (or nucleoids) with exogenous nucleases followed by separation of matrix-bound and soluble DNA fractions. It is known that large pieces of chromatin are not soluble in physiological ionic strength. Hence, extractions with concentrated salt solutions were commonly used for disruption of chromatin and elution of cleaved-off DNA (for examples see refs. 1-6). It was proposed later that such extractions may cause different artefacts and hence should be substituted by extractions with a weak ionic detergent, lithium diiodosalicylate (LIS).[7] However, the above extractions do not release all DNA from the nuclei only if the nuclei have been stabilized either with Cu^{2+} ions or by incubation at 37°C (for a review see ref. 8). Furthermore, it turned out that complexes of DNA with nuclear remnants isolated by LIS extraction were not stable upon reextraction with the same solution. Hence, it is most likely that they originate as a result of in vitro reconstitution at the step of treatment of LIS-extracted nuclei with restriction enzymes.[9-11] We shall discuss this procedure along with other studies of reconstituted complexes of DNA with the nuclear matrix or nuclear matrix proteins (chapter 6).

In order to avoid high-salt extraction it was also proposed to separate cleaved-off pieces of DNA arranged in chromatin by electroelution in physiological ionic strength.[12-14] Cells embedded into beads of low-melting agarose are first treated with nucleases and then placed in slots of agarose gel. This is followed by the separation of cleaved-off pieces of DNA loops by electrophoresis. DNA fragments bound to the nuclear matrix (in this case called nucleoskeleton) remain at the start of the gel. As shall be discussed below, the results obtained with this procedure of isolation of matrix-bound DNA fragments fully support the conclusions made in experiments with high-salt extraction.

Recent development of the technique of multicolor fluorescence in situ hybridization (FISH) opens up possibilities for studying the distribution of unique DNA sequences in DNA loops by direct hybridization of fluorescent probes with a halo of nuclear or chromosomal DNA.[15,16] Although the resolution of this approach at present is not very high, the method has an obvious advantage over the isolation and characterization of nuclear matrix DNA as it excludes the possibility of an artificial redistribution of matrix-bound DNA sequences at the step of nuclease treatment and subsequent separation of matrix-bound and cleaved-off DNA.

4.2. A COMPLEX PATTERN OF DNA INTERACTIONS WITH THE NUCLEAR MATRIX

4.2.1. Association of replicating DNA sequences with the nuclear matrix

Already, in initial experiments on nuclear matrix isolation, it was found that nuclear matrix DNA was highly enriched in replicating sequences.[17] After a short incubation of cells in the presence of radiolabeled precursors for DNA synthesis, the greater part of radiolabeled DNA could be found in the matrix-bound fraction. With an increase of the pulse time or after a chase in the medium supplemented with "cold" precursors, the labeled DNA was most entirely recovered in the supernatant, i.e., it was eluted from the nuclear matrix by combined low salt and high salt extractions (Fig. 4.1).[17,18] Similar observations were made later by other authors.[19-26] Especially persuasive observations were, perhaps, made in experiments with autoradiographic visualization of nascent DNA in nucleoid loops.[27,28] After a short pulse the whole radioactive label resided within the confines of the nuclear matrix. With an increase in the labeling time it was possible to observe the migration of the labeled DNA into the halo of decondensed DNA loops. Using this approach, the authors calculated an approximate speed of movement of the labeled DNA from the nuclear matrix to the distal parts of DNA loops, which must reflect the speed of progression of replication forks along the DNA template.[27,28] The obtained value (1 µm/min) agreed well with other estimates of the speed of movement of the replication forks.

Replicating DNA possesses some properties which distinguish it from the bulk DNA. For example, there are branched and single-stranded DNA regions at or in the vicinity of the replication forks. Many nuclear matrix proteins are able to bind selectively single-stranded DNA (see chapter 6). Hence, the possibility should be considered that the apparent association of replicating DNA with the nuclear matrix is an artefact reflecting preferential detaining of replicating DNA by the nuclear matrix proteins in the course of nuclear matrix isolation. Special control experiments aimed to check

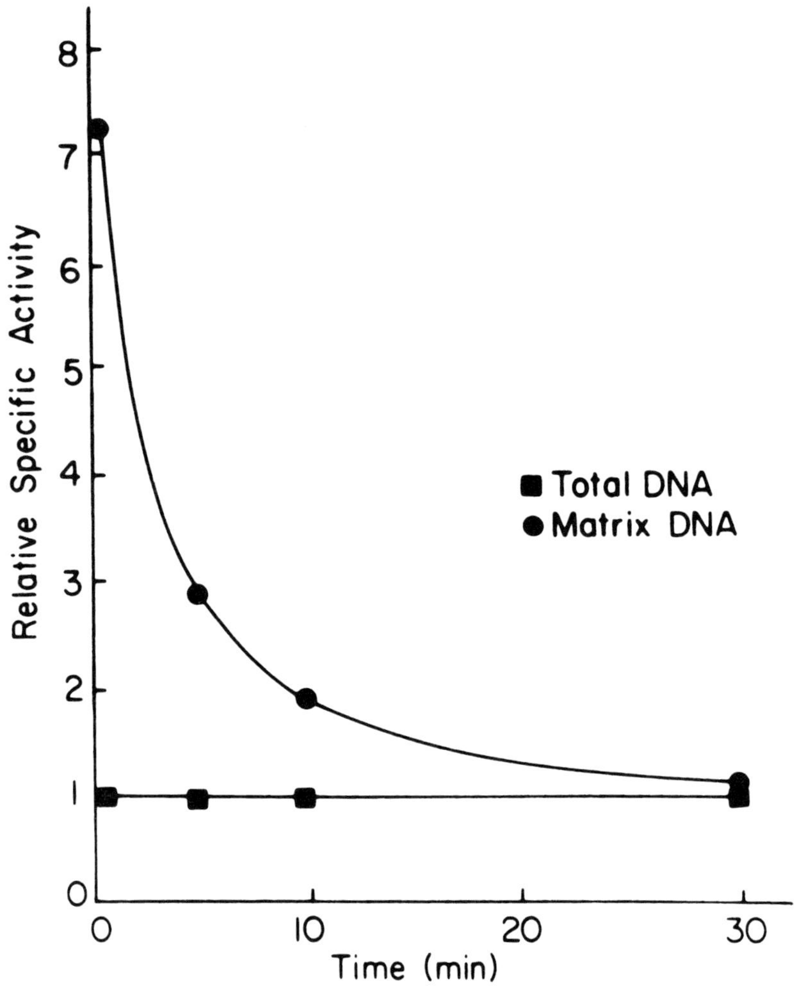

Fig. 4.1. Kinetics of association of newly replicated DNA with the nuclear matrix of regenerating rat liver. Regenerating rat livers were labeled for various times after injection of 200 μCi ^{3}H-thymidine into the portal vein of partially hepatectomized rats. Specific activities of the matrix DNA relative to total DNA are shown for various pulse times. (Reprinted with permission of Cell Press from Pardoll DM, Vogelstein B, Coffey DS. Cell 1980; 19:527-536. © 1980 by Cell Press).

this possibility were carried out by Berezney and Buchholtz.[18] It was demonstrated that exogenous replicating DNA or replicating chromatin mixed with nuclei or nuclear matrices at different steps of their isolation were not preferentially detained by the nuclear matrices. These data support the conclusion that interactions of the endogenous replicating DNA with the nuclear matrix may not originate in the course of nuclear matrix isolation, but rather preexist in living cells. The supposition that replication occurs at the nuclear matrix is further supported by the demonstration that replication forks are concentrated in the matrix-bound DNA fraction.[29,30] Finally, it was found that an isolated nuclear matrix could continue DNA replication initiated in living cells.[31-33]

4.2.2. Association of transcriptionally active DNA with the nuclear matrix

In a number of experiments, a relationship between the transcriptional status of a gene and its attachment to the nuclear matrix was observed. Active genes were generally found in the nuclear matrix DNA, while nonactive ones—in cleaved-off DNA fractions.[2-6,34-38] Especially convincing results were obtained in experiments with hormone-induced premature cell differentiation. Cell type-specific genes starting their expression in differentiated cells were bound to the nuclear matrix in these cells and were not bound to the nuclear matrix in maternal (nondifferentiated) cells.[4,35,36] For example, it was shown that the vitellogenin gene was not associated with the nuclear matrix in liver cells of immature chicks. However, after primary stimulation of premature differentiation by estradiol treatment this gene was found mainly in the matrix-bound DNA fraction.[35] Concomitantly, although not exactly simultaneously, a sharp increase in the level of vitellogenin messenger RNA was detected. However, in the absence of secondary stimulation the rate of vitellogenin gene transcription gradually decreased and finally returned to the basal level. Fifteen days after the stimulation vitellogenin mRNA was barely detectable. The decrease in the level of vitellogenin gene transcription correlated with an apparent dissociation of this gene from the nuclear matrix.[35] Other evidence also supports the conclusion that active genes are for the most part bound to the nuclear matrix. Thus, a clear relationship between the expression of viral genomes integrated into the host cell DNA and their attachment to the nuclear matrix was observed.[2,39]

It is obvious that these results were far from what was expected. No wonder that many scientists treated them with a certain suspicion. It was proposed that the apparent association of transcriptionally active DNA with the nuclear matrix was an artefact of high-salt extraction.[7,40,41] Indeed, it is known that RNA can be precipitated by a concentrated salt solution. The efficiency of precipitation depends to a certain extent on the concentration of RNA. The local concentration of RNA (including nascent RNA bound to DNA through transcriptional complexes) is very high in eukaryotic cell nuclei. Hence, the possibility of RNA precipitation in the course of treatment of nuclei with 2M NaCl solution cannot be ruled out.

Several lines of experimental evidence show, however, that precipitation of either nascent RNA or transcriptional complexes is not the cause (or, to be more precise, is not the only cause) of detainment of active genes in the nuclear matrix. First, neither the organization of DNA into loops (in nucleoids), nor the association of active genes with the nuclear matrix are affected by RNase treatment.[42-44] Second, the inhibition of transcription by different drugs does not cause the detachment of active genes from the nuclear matrix.[6,44] Third, the arrest of transcription of most of the normally active genes under heat shock conditions (and the dissociation of transcriptional complexes from these genes) also does not cause the detachment from the nuclear matrix of normally transcribed, although temporarily silent, genes.[6] Fourth, in the course of hormone-induced cell differentiation an association of cell type-specific genes with the nuclear matrix occurs before the actual start of transcription.[35] Fifth, all members of the ovalbumin gene family are equally represented in the oviduct cells nuclear matrix DNA in spite of quite different rates of their transcription.[34] Sixth, an apparent association of active genes with the nuclear matrix was also observed in experiments with isolation of nuclear matrix-bound DNA fragments under isotonic conditions[12,45,46] unless the nuclei were preincubated in a low ionic strength solution.[12] The effect of low-salt treatment may be due to the dissociation under these conditions of some nuclear matrix structures including actin filaments (see chapter 1). It should be mentioned that preincubation of nuclei in a solution with a very low ionic strength also prevents the possibility of observing the association of active genes with nuclear matrices isolated by subsequent high-salt extraction.[47]

Taken together, all the above presented data support the conclusion that transcription per se is not essential for the association of active genes with the nuclear matrix. This association is likely to be a condition of transcriptional activation, rather than a consequence of transcription. On the basis of recent findings,[48-50] one can suggest that the association of active genes with the nuclear matrix reflects their targeting to the transcription foci.

4.3. IDENTIFICATION OF PERMANENT SITES OF DNA ANCHORAGE ON THE NUCLEAR MATRIX

The functional dependence and heterogeneity of DNA-nuclear matrix interactions could hardly be explained by the simple model of DNA organization into relatively uniform loops anchored to the nuclear matrix. On the other hand, the assumption that all specific DNA interactions with the nuclear matrix are transcription-related[5,51,52] also failed to explain some experimental observations—for example, the correlation between an average size of DNA loops and an average size of replicons in cells of different species,[53] the possibility of preparing nucleoids with constrained DNA loops from transcriptionally silent mature sperm nucleiz[54] or the similarity of apparent loop sizes in euchromatic (i.e., transcribed and potentially transcribed) and heterochromatic (i.e., nontranscribed) regions of chromosomes.[16] Unique DNA sequences are distributed nonrandomly versus DNA loop anchorage sites in both mature sperm nuclei[55] and mature avian erythrocyte nuclei (see below and also refs. 56, 57). To elucidate the problem, the existence was suggested along with function-dependent (transient) DNA-nuclear matrix associations of "structural" or "permanent" attachment sites which delimit the borders of supercoiled DNA loops in nucleoids and which are not related to transcription. In other words, it was proposed to correct the initial assumption by supposing that only a part (and may be not the major one) of nuclear matrix DNA fragments represents the basements of topological DNA loops.[56-59] If this supposition is correct, transcription-related and other transient associations of DNA with the nuclear matrix may be responsible for the anchorage of the majority of nuclear matrix DNA fragments without affecting the average size of DNA loops estimated by the nucleoid relaxation analysis. As has been discussed before (chapter 3), the sedimentation properties of nucleoids are determined by the status of large DNA loops (i.e., those loops which are not partitioned into sub-domains by transient interactions of DNA with the nuclear matrix). The diameter of the DNA loop halo in nucleoids also reflects the size of large loops.

Large loops subdivided into smaller ones by any type of function-dependent interactions of DNA with the nuclear matrix are likely to be hidden within the nucleiod core and even after relaxation these loops will be smaller than the halo diameter (for a discussion see ref. 10).

In an attempt to discriminate the basic loop anchorage sites from the functional interactions of DNA with the nuclear matrix we have mapped the preferential positions of DNA attachment to the nuclear matrix in the domain of chicken α–globin genes in erythroblasts, mature erythrocytes and mature sperm nuclei.[56,57,60] The rationale for this experiment was that in terminally-differentiated transcriptionally-silent mature erythrocytes and mature sperm nuclei, transcription-dependent associations of DNA with the nuclear matrix might disappear while permanent attachment sites (if any) were likely to be preserved. Indeed, while in erythroblast nuclei the whole transcriptionally active area was bound to the nuclear matrix, in erythrocyte and sperm nuclei only two major attachment sites framing the domain from the upstream and the downstream were detected (Fig. 4.2). The same DNA fragments were also attached to the nuclear matrix of cultured chicken fibroblasts.[60] Hence, at least within the domain of chicken α-globin genes, it was possible to discriminate permanent (preserved in nonactive nuclei) and transcription-dependent interactions of DNA

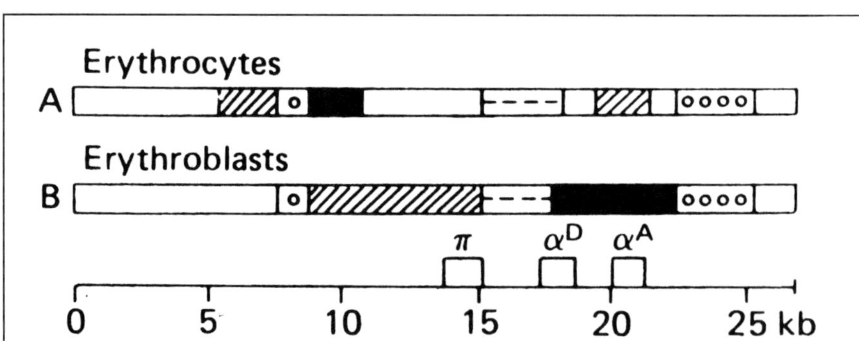

Fig. 4.2. Preferential positions of DNA attachments to the nuclear matrix within the domain of chicken α-globin genes in erythrocyte (A) and erythroblast (B) nuclei. The regions highly and moderately enriched in nuclear matrix DNA are shown by dark color and by sloping lines, respectively. Circles indicate the restriction fragments including repetitive sequences which are abundant both in total and in nuclear matrix DNA. Nontested area is marked by a broken line. The positions of globin genes are shown by open rectangles above the distance scale. (Reprinted with permission of ICSU Press from Razin SV. BioEssays 1987; 6:19-23. © 1987 ICSU Press.)

with the nuclear matrix. It is important to mention that in the above experiments the nuclear matrices were reextracted with 2M NaCl solution after nuclease treatment.[56,57,60] Hence, permanent DNA-nuclear matrix interactions found in erythrocyte and sperm nuclei could not be artificially caused by rebinding of some DNA sequences to the affinity sites on an isolated nuclear matrix. Incubation of erythrocyte nuclei with a nuclease was carried out at 37°C. Although made with the aim to introduce cleavages into loop DNA, this incubation also ensured stabilization of the internal nuclear matrix permitting the preservation of its integrity during high-salt extraction (see refs. 61, 62). Nuclear matrices isolated from mature chicken erythrocytes nonstabilized by heating were reported to lack an internal matrix.[63]

An important question is whether permanent sites of DNA attachment to the nuclear matrix observed in the domain of chicken α-globin genes can be found also in other genomic areas. In order to clarify this question, we have compared the sequence complexities of nuclear matrix DNA samples prepared from chicken erythroblasts and mature erythrocytes.[64,65] Trace amounts of either total DNA or nuclear matrix DNA from erythroblasts or erythrocytes were reannealed in reactions driven by a vast excess of either erythrocyte or erythroblast nuclear matrix DNA (Table 4.1). The most

Table 4.1. Comparison of different nuclear matrix DNA preparations with the aid of renaturation experiments

	Percentage of unique sequences co-renatured with the excess of the driver			$C_0t_{\frac{1}{2}}$ value for unique sequences		
Driver... / Labelled probe	Total chicken DNA	Erythroblast nmDNA	Erythrocyte nmDNA	Total chicken DNA	Erythroblast nmDNA	Erythrocyte nmDNA
Total chicken DNA	100	25	10	10^3	$2{\cdot}5 \times 10^2$	10^2
Erythroblast nmDNA	100	100	35	10^3	$2{\cdot}5 \times 10^2$	10^2
Erythrocyte nmDNA	100	100	100	10^3	$2{\cdot}5 \times 10^2$	10^2
oriDNA (erythroblast)	100	100	100	10^3	$2{\cdot}5 \times 10^2$	10^2

Three nick translated DNA probes (i) total chicken DNA, (ii) nuclear matrix DNA from chicken erythroblasts (5 kb-size fragments), (iii) nuclear matrix DNA from mature chicken erythrocytes (5 kb-size fragments) were cleaved to about 200 bp fragments and reannealed with the excess of either total chicken DNA or one of the nuclear matrix DNA preparations. (Reprinted with permission of ICSU Press from Razin SV. BioEssays 1987; 6:19-23. © 1987 ICSU Press.)

important observation made in these experiments was that in both renaturation reactions with C_0t values sufficient for complete renaturation of homologous nuclear matrix DNA probes the total DNA was reannealed only partially (25% of unique sequences in the reaction driven by erythroblast nuclear matrix DNA and 10% of unique sequences in the reaction driven by erythrocyte nuclear matrix DNA). This means that both nuclear matrix DNA fractions contain only subsets of unique DNA sequences present in total DNA. Consequently, one may conclude that matrix attachment sites have specific positions in the genome. Otherwise, all unique sequences present in total DNA would be found in the vicinity of matrix attachment sites (i.e., in nuclear matrix DNA). Similar considerations made it possible to conclude that erythrocyte nuclear matrix DNA is a subfraction of erythroblast nuclear matrix DNA. Indeed, the latter was not reannealed completely in the reaction driven by erythrocyte nuclear matrix DNA while both nuclear matrix DNA samples were reannealed completely in the reaction driven by erythroblast nuclear matrix DNA. Hence, the erythroblast nuclear matrix DNA is composed of two subfractions: DNA fragments permanently attached to the nuclear matrix (i.e., the erythrocyte nuclear matrix DNA) and DNA fragments interacting with the nuclear matrix in accordance with the functional processes.

4.4. CONCLUSIONS

In active nuclei the nuclear matrix DNA is composed of two subfractions, namely of DNA fragments permanently attached to the nuclear matrix (which most likely are the basements of topological DNA loops) and fragments transiently bound to the nuclear matrix due to different functional processes. DNA fragments permanently attached to the nuclear matrix constitute 30% or even less of the whole nuclear matrix DNA. In nonactive nuclei (avian erythrocytes or mature sperm nuclei) the whole nuclear matrix DNA seems to be composed of the permanently attached DNA fragments.

Related Review Articles

Cook PR. The nucleoskeleton: artefact, passive network or active site? J Cell Sci 1988; 90:1-6.

Getzenberg PH, Pienta KJ, Ward WS et al. Nuclear structure and three-dimensional organization of DNA. J Cell Biochem 1991; 4:289-299.

Jackson DA, McCready SJ, Cook PR. Replication and transcription depend on attachment of DNA to the nuclear cage. J Cell Sci 1984; Suppl. 1:59-79.

Jackson DA. Structure-function relationship in eukaryotic cell nuclei. Bioessays 1991; 13:1-10.

Jackson DA, Dolle A, Robertson G et al. The attachments of chromatin loops to the nucleoskeleton. Cell Biol Int Rep 1992; 16:687-696.

Razin SV. DNA interactions with the nuclear matrix and spatial organization of replication and transcription. Bioessays 1987; 6:19-23.

Razin SV, Vassetzky YS. Domain organization of the eukaryotic genome. Cell Biol Int Rep 1992; 16:697-708.

Zlatanova JS, Van Holde KE. Chromatin lops and transcription regulation. Crit Rev Eukar Gene Expr 1992; 2:211-224.

Cook PR. RNA polymerase: structural determinant of the chromatin loop and chromosome. Bioessays 1994; 16:425-430.

References

1. Cook PR, Brazell IA. Mapping sequences in loops of nuclear DNA by their progressive detachment from the nuclear cage. Nucl Acids Res 1980; 8:2895-2906.
2. Cook PR, Lang J, Hayday et al. Active viral genes in transformed cells lie close to the nuclear cage. EMBO J 1982; 1:447-452.
3. Robinson SI, Nelkin BD, Vogelstein B. The ovalbumin gene is associated with the nuclear matrix of chicken oviduct cells. Cell 1982; 28:99-106.
4. Robinson SI, Small D, Idzerda R et al. The association of active genes with the nuclear matrix of the chicken oviduct. Nucl Acids Res 1983; 15:5113-5130.
5. Small D, Vogelstein B. The anatomy of supercoiled loops in the *Drosophila* 7F locus. Nucl Acids Res 1985; 13:7703-7713.
6. Small D, Nelkin B, Vogelstein B. The association of transcribed genes with the nuclear matrix of *Drosophila* cells during heat shock. Nucl Acids Res 1985; 13:2413-2431.
7. Mirkovich J, Mirault E, Laemmli UK. Organization of the higher order chromatin loop: specific DNA attachment sites on nuclear scaffold. Cell 1984; 39:323-332.
8. Gasser SM, Amati BB, Cardenas ME et al. Studies on scaffold attachment sites and their relation to genome function. Int Rev Cytol 1989; 119:57-96.
9. Izaurralde E, Mirkovich J, Laemmli UK. Interaction of DNA with nuclear scaffolds in vitro. J Mol Biol 1988; 200:111-125.
10. Razin SV, Gromova II, Iarovaia OV. Specificity and functional significance of DNA interaction with the nuclear matrix: New approaches to clarify the old questions. International Rev Cytol 1995; 162B:405-448.

11. Razin SV. Functional architecture of chromosomal DNA domains. Crit Rev Eucaryot Genes Exp 1996; (in press).
12. Jackson DA, Yuan J, Cook PR. A gentle method for preparing cyto- and nucleo-skeletons and associated chromatin. J Cell Sci 1988; 90:365-378.
13. Jackson DA, Cook PR. Transcription occurs at a nucleoskeleton. EMBO J 1985; 4:919-925.
14. Jackson DA, Cook PR. Replication occurs at a nucleoskeleton. EMBO J 1986; 5:1403-1410.
15. Gerdes MG, Carter KC, Moen PT et al. Dynamic changes in the higher-level chromatin organization of specific sequences revealed by in situ hybridization to nuclear halos. J Cell Biol 1994; 126:289-304.
16. Bickmore WA, Oghene K. Visualizing the spatial relationships between defined DNA sequences and the axial region of extracted metaphase chromosomes. Cell 1996; 84:95-104.
17. Berezney R, Coffey DS. Nuclear protein matrix: association with newly synthesized DNA. Science 1975; 189:291-292.
18. Berezney R, Buchholtz LA. Dynamic association of replicating DNA fragments with the nuclear matrix of regenerating liver. Exp Cell Res 1981; 132:1-13.
19. Dijkwel PA, Mullenders LH, Wanka F. Analysis of the attachment of replicating DNA to a nuclear matrix in mammalian interphase nuclei. Nucl Acids Res 1979; 6:219-230.
20. Pardoll DM, Vogelstein B, Coffey DS. A fixed site of DNA replication in eukaryotic cells. Cell 1980; 19:527-536.
21. Hunt BF, Vogelstein B. Association of newly replicated DNA with the nuclear matrix of Physarum polycephalum. Nucl Acids Res 1981; 9:349-363.
22. Aelen JM, Opstelsten RJ, Wanka F. Organization of DNA replication in Physarum polycephalum. Attachment of origins of replication and replication forks to the nuclear matrix. Nucl Acids Res 1983; 11:1181-1195.
23. Konstantinovic M, Sevaljevic L. Nuclear matrix from resting and concanavalin A-stimulated human lymphocytes. Biochim Biophys Acta 1983; 762:1-8.
24. Van der Velden HM, Poot M, Wanka F. In vitro DNA replication in association with the nuclear matrix of permeable mammalian cells. Biochim Biophys Acta 1984; 782:429-436.
25. Conrad MN, Zakian VA. Nuclear matrix involvement in DNA replication in Saccharomyces cerevisiae. J Cell Biol 1984; 99 Pt2:127.
26. Fernandes DJ, Smith-Nanni C, Paff MT et al. Effects of antileukemic agents on nuclear matrix-bound DNA replication in CCRF-CEM leukemia cell. Cancer Res 1989; 48:1850-1855.
27. Vogelstein P, Pardoll DM, Coffey DS. Supercoiled loops and eukaryotic DNA replication. Cell 1980; 22:79-85.

28. McCready SJ, Godwin J, Mason DW et al. DNA is replicated at the nuclear cage. J Cell Sci 1980; 46:365-386.
29. Valenzuela MS, Mueller GC, Dasgupta S. Nuclear matrix DNA complex resulting from EcoRI digestion of HeLa nucleoids is enriched for DNA replication forks. Nucl Acids Res 1983; 11: 2155-2164.
30. Vaughn JP, Dijkwel PA, Mullenders LHF et al. Replication forks are associated with the nuclear matrix. Nucl Acids Res 1990; 18:1965-1969.
31. Smith HC, Berezney R. Nuclear matrix-bound deoxyribonucleic acid synthesis: an in vitro system. Biochemistry 1982; 21:6751-6761.
32. Tubo RA, Smith HC, Berezney R. The nuclear matrix continues DNA synthesis at in vivo replication forks. Biochim Biophys Acta 1985; 825:326-334.
33. Shioda M, Matsuzawa Y, Murakami-Murofushi K et al. DNA synthesis by the isolated nuclear matrix from synchronized plasmodia of Physarum polycefalum. Biochim Biophys Acta 1989; 1007: 254-263.
34. Ciejek EM, Tsai M-J, O'Malley BW. Actively transcribed genes are associated with the nuclear matrix. Nature 1983; 306:607-609.
35. Jost J-P, Seldran M. Association of transcriptionally active vitellogenin II gene with the nuclear matrix of chicken liver. EMBO J 1984; 3:2205-2208.
36. Mirell CJ, Bekhor I. The effect of estrogen on the concentration of ovalbumin gene sequence in the 2M NaCl residual fraction of oviduct chromatin. FEBS Lett 1982; 150:117-121.
37. Hentzen PC, Rho JH, Bekhor I. Nuclear matrix DNA from chicken erythrocytes contains β-globin gene sequences. Proc Natl Acad Sci USA 1984; 81:304-307.
38. Dalton S, Younghusband BH, Wells JRE. Chicken histone genes retain nuclear matrix association throughout the cell cycle. Nucl Acids Res 1986; 14:6507-6523.
39. Nelkin BD, Pardoll D, Vogelstein B. Localization of SV40 genes within supercoiled loop domains. Nucl Acids Res 1980; 8: 5623-5633.
40. Kirov N, Djondjurov L, Tsanev R. Nuclear matrix and transcriptional activity of the mouse α-globin gene. J Mol Biol 1984; 180:601-614.
41. Roberge M, Dahmus ME, Bradbury EM Chromosomal loop/nuclear matrix organization of transcriptionally active and inactive RNA polymerases in HeLa nuclei. J Mol Biol 1988; 201:545-555.
42. Cook P, Brazell IA. Spectro fluorometric measurement of the binding of ethidium bromide to superhelical DNA from cell nuclei. Eur J Biochem 1978; 84:464-477.
43. Vogelstein B, Small D, Robinson S et al. The nuclear matrix and the organization of nuclear DNA. In: Bekhor I, ed. Progress in

Nonhistone Protein Research. Boca Raton, FL: CRC Press, Inc., 1985; 2:115-129.

44. Iarovaia OV, Gorelova TV, Shuppe NG et al. A pattern of the heat shock genes attachment to the nuclear matrix does not depend on their functional state. Genetica (Moscow) 1985; 21: 896-901.
45. Thorburn A, Moore R, Knowland J. Attachment of transcriptionally active DNA sequences to the nucleoskeleton under isotonic conditions Nucl Acids Res 1988; 16:7183.
46. Thorburn A, Knowland J. Attachment of vitellogenin genes to the nucleoskeleton accompanies their activation. Biochem Biophys Res Commun 1993; 191:308-313.
47. Razin SV, Yarovaia OV, Georgiev GP. Low ionic strength extraction of nuclease-treated nuclei destroys the attachment of transcriptionally active DNA to the nuclear skeleton. Nucl Acids Res 1985; 13:7427-7444.
48. Jackson DA, Hassan AB, Errington RJ et al. Visualization of focal sites of transcription within human nuclei. EMBO J 1993; 12:1059-1065.
49. Xing Y, Johnson CV, Dobner PR et al. Higher level organization of individual gene transcription and RNA splicing. Science 1993; 259:1326-1330.
50. Xing Y, Johnson CV, Moen Jr PT et al. Nonrandom gene organization: Structural arrangements of specific pre-mRNA transcription and splicing with SC-35 domains. J Cell Biol 1995; 131:1635-1647.
51. Cook PR. RNA polymerase: structural determinant of the chromatin loop and chromosome. Bioessays 1994; 16:425-430.
52. Cook PR. A chromomeric model for nuclear and chromosome structure. J Cell Sci 1995; 108:2927-2935.
53. Buongiorno-Nardelli M, Gioacchino M, Carri MT et al. A relationship between replicon size and supercoiled loop domains in the eukaryotic genome. Nature 1982; 298:100-102.
54. Ward WS, Partin AW, Coffey DS. DNA loop domains in mammalian spermatozoa. Chromosoma 1989; 98:153-159.
55. Ward WS, Coffey DS. Specific organization of genes in relation to the sperm nuclear matrix. Biochem Biophys Res Commun 1990; 173:20-25.
56. Razin SV, Rzeszowska-Wolny J, Moreau J et al. Localization of regions of DNA attachment to the nuclear skeleton within chicken alpha-globin genes in functionally active and functionally inactive nuclei. Moltculiarnaia Biologia (Moscow) 1985; 19:456-466.
57. Farache G, Razin SV, Rzeszowska-Wolny J et al. Mapping of structural and transcription-related matrix attachment sites in the α-globin gene domain of avian erythroblasts and erythrocytes. Mol Cell Biol 1990; 10:5349-5358.

58. Pienta KJ, Getzenberg RH, Coffey DS. Cell structure and DNA organization. Crit Rev Eukar Gene Expres 1991; 1:355-385.
59. Getzenberg RH, Pienta KJ, Ward WS et al. Nuclear structure and the three-dimensional organization of DNA. J Cell Biochem 1991; 47:289-299.
60. Kalandadze AG, Bushara SA, Vassetzky YS et al. Characterization of DNA pattern in the site of permanent attachment to the nuclear matrix located in the vicinity of replication origin. Biochem Biophys Res Commun 1990; 168:9-15.
61. Hendzel MJ, Davie JR. Nuclear distribution of histone deacetylase: A marker enzyme for the internal nuclear matrix. Biochim Biophys Acta 1992; 1130:307-313.
62. Hendzel MJ, Sun J-M, Chen HY et al. Histone acetyltransferase is associated with the nuclear matrix. J Biol Chem 1994; 269: 22894-22901.
63. Lafond RE, Woodcock CLF. Status of the nuclear matrix in mature and embryonic chick erythrocyte nuclei. Exp Cell Res 1983; 147:31-39.
64. Razin SV, Kekelidze MG, Lukanidin EM et al. Replication origins are attached to the nuclear skeleton. Nucl Acids Res 1986; 14:8189-8207.
65. Razin SV. DNA interactions with the nuclear matrix and spatial organization of replication and transcription. Bioessays 1987; 6:19-23.

CHAPTER 5

Chromosomal DNA Loops can be Excised from the Genome by Topoisomerase II-Mediated DNA Cleavage at Matrix Attachment Sites

5.1. DESCRIPTION OF AN APPROACH FOR TOPOISOMERASE II-MEDIATED EXCISION OF CHROMOSOMAL DNA LOOPS

A new approach for characterizing the specificity of chromosomal DNA organization into loops has been developed recently in our laboratory in collaboration with the laboratory of Prof. Ronald Hancock (Laval University Cancer Research Centre, Quebec, Canada). The conceptual difference of this new approach from all previously used ones is that instead of isolating DNA sequences located close to the loop basements (nuclear matrix DNA) it is proposed to characterize individual DNA loops after their excision by targeted DNA cleavage at matrix attachment sites. It seems that the nature itself has supplied investigators with an instrument necessary for the excision of chromosomal DNA loops. This instrument is an abundant enzyme, DNA topoisomerase II. As mentioned before (chapters 1 and 2), DNA topoisomerase II is an important component of the protein structures to which DNA loops are bound (i.e., of nuclear matrix and chromosomal scaffold).[1-4] Topoisomerase II is known to induce temporary double-stranded scissions in DNA in the course of a relaxation reaction normally catalyzed by this enzyme[5-7] (Fig. 5.1). Intermediate cleavage complexes can be accumulated by inhibition of the religation step of the topoisomerase II-mediated cleavage-reunion reaction with a variety of antitumor drugs, such as VM-26, VP-16 or

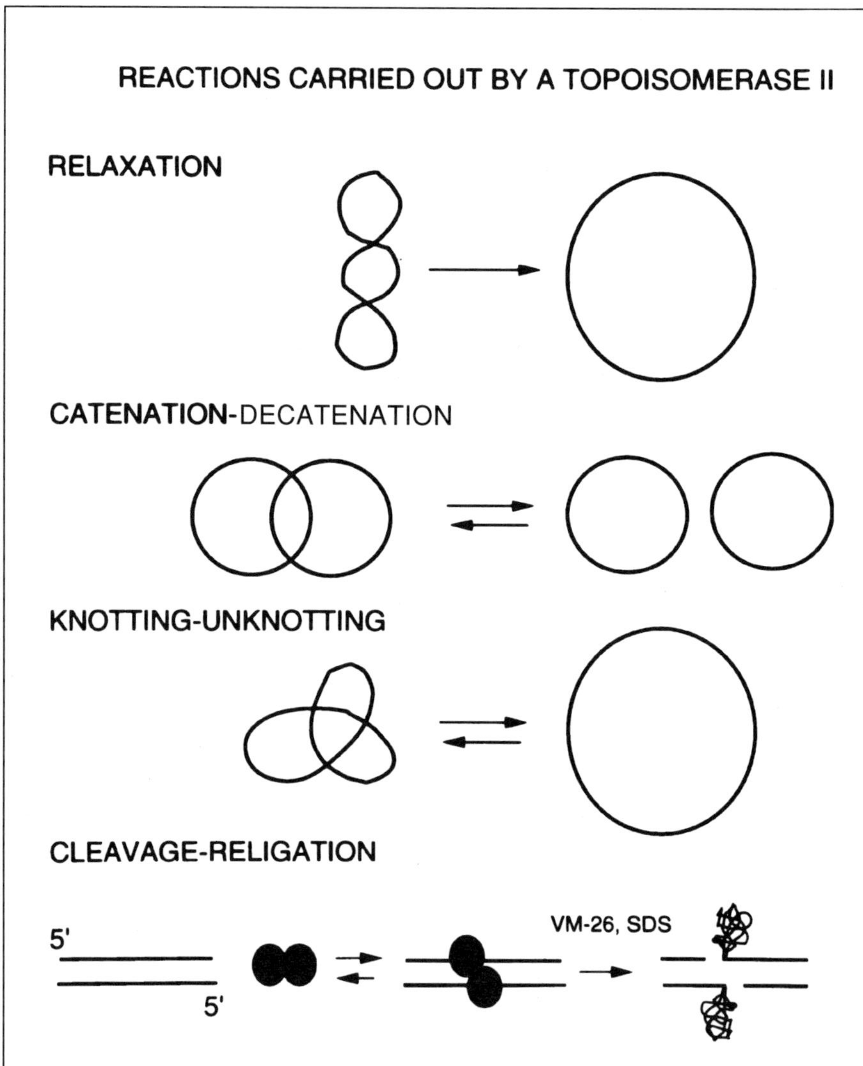

Fig. 5.1. A scheme illustrating different reactions catalyzed by DNA topoisomerase II.

m-AMSA.[5,6,8-10] Digestion of the protein moiety of topoisomerase II-DNA complexes by proteinase K reveals double stranded breaks at the sites of topoisomerase II action on DNA (Fig. 5.1). It has been found previously that topoisomerase II-mediated cleavage of free DNA is characterized by a limited sequence specificity.[11-14] In living cells, the accessible places in chromatin were found to be preferential targets for topoisomerase II-mediated cleavage.[15-17] It should be mentioned that two isozymes of mammalian DNA topoisomerase II (topoisomerase II α, 170 kDa and topoisomerase II β, 180 kDa) have been identified.[18] These isozymes are encoded by two different genes.[19] Topoisomerase II α is a major and more stable isozyme. Hence, most of the previous observations related to the properties of eukaryotic topoisomerase II actually reflect the properties of this isozyme. Topoisomerase II β is not present in the nuclear matrix, while topoisomerase II α is partitioned between the nuclear matrix (high-salt-insoluble nuclear compartment) and soluble nuclear compartment.[20-24] A significant portion of this enzyme can be extracted from nuclei using a medium-salt solution.[20-24] It was reasonable to assume that soluble topoisomerases II of nucleoplasm may have preferential access to DNase I hypersensitive sites in chromatin. It is known (see chapters 1 and 2) that the extraction of nuclei with concentrated salt solutions does not affect the nuclear matrix (chromosomal scaffold) integrity and does not release DNA loops from the nuclear matrix. At the same time, this extraction may be expected to remove soluble topoisomerases II able to interact with DNA loops outside the matrix attachment regions. Furthermore, the preferential accessibility of DNase I hypersensitive regions for soluble topoisomerase II must be lost in any case with the disruption of nucleosomes by 2M NaCl extraction. It is also important that topoisomerase II α is not inactivated by incubation in a concentrated salt solution. As a matter of fact, the high-salt-insoluble topoisomerase II of the nuclear matrix was shown to be able to introduce scissions in DNA.[25]

On the basis of the above facts we have elaborated the following protocol for topoisomerase II-mediated DNA loop excision and mapping the DNA loop organization in large genomic areas:

Cells were permeabilized with a nonionic detergent and extracted with 2M NaCl in order to obtain nucleoids,[26] i.e., the remnants of nuclei with extended DNA loops attached to the nuclear matrix. In some initial experiments, the nuclear matrix was stabilized with Cu^{2+} ions before high-salt extraction. However, this step was found to be not necessary. Furthermore, Cu^{2+} ions were demonstrated to inhibit topoisomerase II-mediated DNA cleavage (Gromova II and Razin SV, unpublished results).

After being washed from the high-salt-extraction buffer, the nucleoids were incubated under conditions favoring topoisomerase II-mediated DNA cleavage (in a low-salt, magnesium-containing buffer supplemented with a topoisomerase II-specific drug, VM-26). After termination of the incubation by addition of SDS and digestion of proteins with proteinase K, the excised loops and their oligomers were separated by pulsed field gel electrophoresis (PFGE) and subjected to Southern analysis.

It is hardly possible to preserve the integrity of high-molecular weight DNA during high-salt extraction and subsequent transfer to the buffer for topoisomerase II-mediated DNA cleavage. Besides, for Southern analysis, the DNA should be loaded on gels for PFGE at relatively high concentrations. To solve both problems we have carried out all extractions and subsequent treatments after embedding cells into thin agarose blocks.

5.2. STUDY OF CHROMOSOMAL DNA LOOP ORGANIZATION IN A CLUSTER OF MAMMALIAN RIBOSOMAL GENES

The above described protocol for the excision of chromosomal DNA loops was first tested in experiments with mammalian ribosomal genes.[27,28] It was found that these genes were organized into loops equal in size to an individual repeated unit and separated by the attachment sites in nontranscribed spacers. Indeed, when the products of DNA cleavage by endogenous topoisomerase II in high-salt-extracted nuclei were separated by PFGE and subjected to Southern analysis with a ribosomal DNA (rDNA) probe, multiple bands with the sizes divisible by the size of an individual ribosomal gene repeat were observed (Fig. 5.2). This fragmentation pattern could not be reproduced when purified DNA was treated with exogenous topoisomerase II and hence was not due to the limited sequence specificity of DNA cleavage by topoisomerase II. At high VM-26 concentrations virtually all nucleolar DNA was cut to fragments having the size of an individual rDNA repeat (Fig. 5.2). This fragmentation pattern was not altered in mitotic

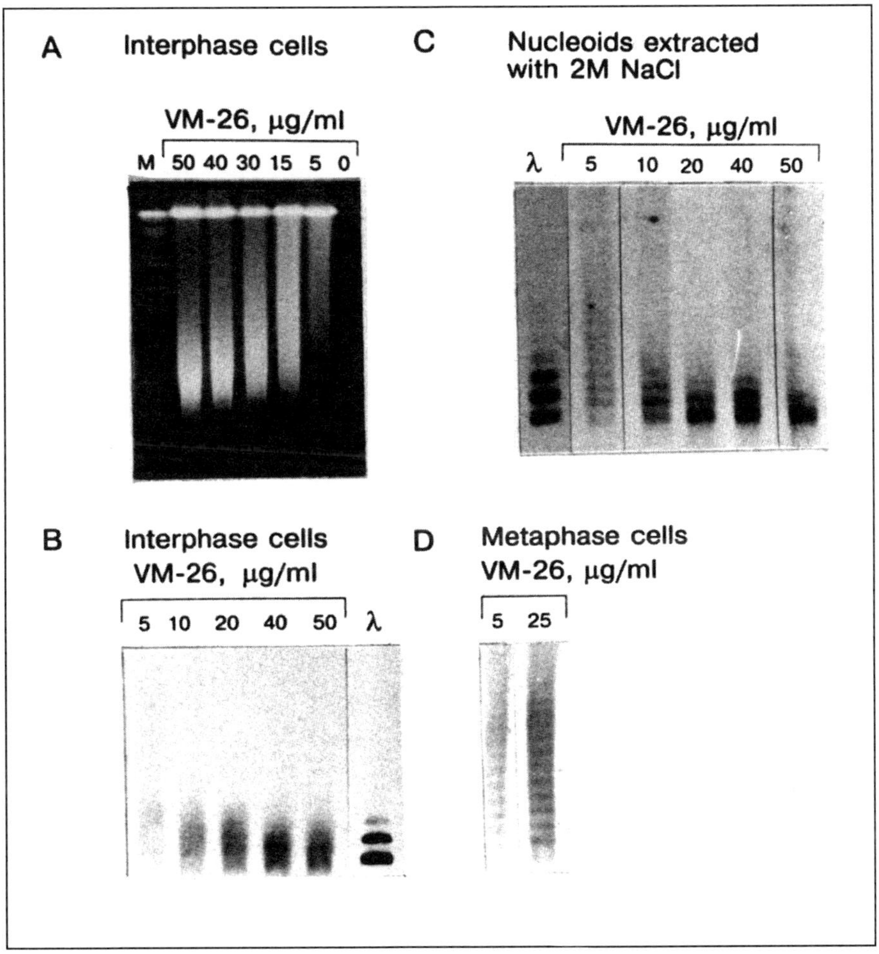

*Fig. 5.2. Cleavage of the Chinese hamster cultured cell (CHO cell line) ribosomal genes by an endogenous topoisomerase II in living cells and in high-salt-extracted nuclei. (A) Size distribution of DNA fragments released from the genome of CHO cells incubated in the media containing different amounts of VM-26 (staining with ethidium bromide after separation by PFGE). Cells embedded in agarose blocks were incubated with 0-50 µg/ml of VM-26 as indicated above the slots. After cell lysis and digestion of proteins, the released DNA fragments were separated by PFGE. Line "M" shows distribution of a molecular weight marker (*Saccharomyces cerevisiae* chromosomal DNA). (B) After separation by PFGE (see panel "A") the released DNA fragments were transferred to nylon filter and hybridized with the ^{32}P-labeled probe representing coding sequences for 18S and 28S ribosomal RNA. The slot designated "λ" shows distribution of the λ phage DNA concatemers. (C) Cells embedded into agarose blocks were first extracted with 2M NaCl in the presence of 0.2% of NP40 and then were incubated in the buffer for topoisomerase II-mediated DNA cleavage supplemented with different amounts of VM-26, as indicated above the slots. After lysis, digestion of proteins and separation of released DNA fragments by PFGE, the Southern analysis was carried out with the same probe as in the experiment shown in section B. (D) The same experiment as this shown in section B, but metaphase cells were treated with VM-26. (Reprinted with permission of Academic Press, Inc., from Razin SV, Gromova II, Iarovaia OV. Int Rev Cytol 1995; 162B:405-448. © 1995 Academic Press, Inc.).*

cells indicating that the specificity of DNA organization into loops is the same in interphase and metaphase chromosomes. Interestingly, the same pattern of bands, although with a higher nonspecific background, was observed when living cells were treated with VM-26.[27] Apparently, matrix attachment sites constitute preferential targets for topoisomerase II-mediated DNA cleavage even in nondisturbed nuclei. That is why the regular pattern could still be seen in the above experiment. The supposition is corroborated by our previous observation that the whole domain of chicken α-globin genes may be excised from the genome as a single loop by topoisomerase II-mediated DNA cleavages at permanent sites of DNA attachment to the nuclear matrix as a result of brief treatment of living cells with inhibitors of topoisomerase II.[29]

The high background observed in experiments with topoisomerase II-mediated DNA loop excision in vivo can be easily explained by a less specific cleavage of DNA with soluble topoisomerase II reported to occur at all accessible places in chromatin such as DNase I HS and even internucleosomal spacers.[15-17] In full accordance with the proposed model (Fig. 5.3), the high-salt extraction eliminated or at least reduced significantly the above background without affecting the regular pattern of the "loop-sized" bands. The possibility of observing the same specific pattern of large-scale fragmentation of nucleolar DNA upon inhibition of topoisomerase II in living cells and in high-salt-extracted nuclei argues against the supposition that the high-salt treatment causes disintegration, redistribution or creation de novo of matrix attachment sites organized with the participation of topoisomerase II. However, a local redistribution of the sites of DNA contact with topoisomerase II within a matrix attachment region cannot be excluded in experiments made at the level of PFGE resolution.

A high-resolution analysis performed using indirect end-labeling of topoisomerase II cleavage products additionally cut by restriction enzymes made it possible to conclude that both in living cells and in high-salt-extracted nuclei topoisomerase II-mediated cleavages occurred in a nontranscribed spacer within an about 4 kb area located just upstream to the start of the 45S rRNA precursor.[27,28] Hence, at least in this case it would be appropriate to speak about a DNA loop anchorage region (LAR) rather than about a DNA loop anchorage site. The distribution of DNA loop anchorage sites in the human ribosomal gene cluster has recently been

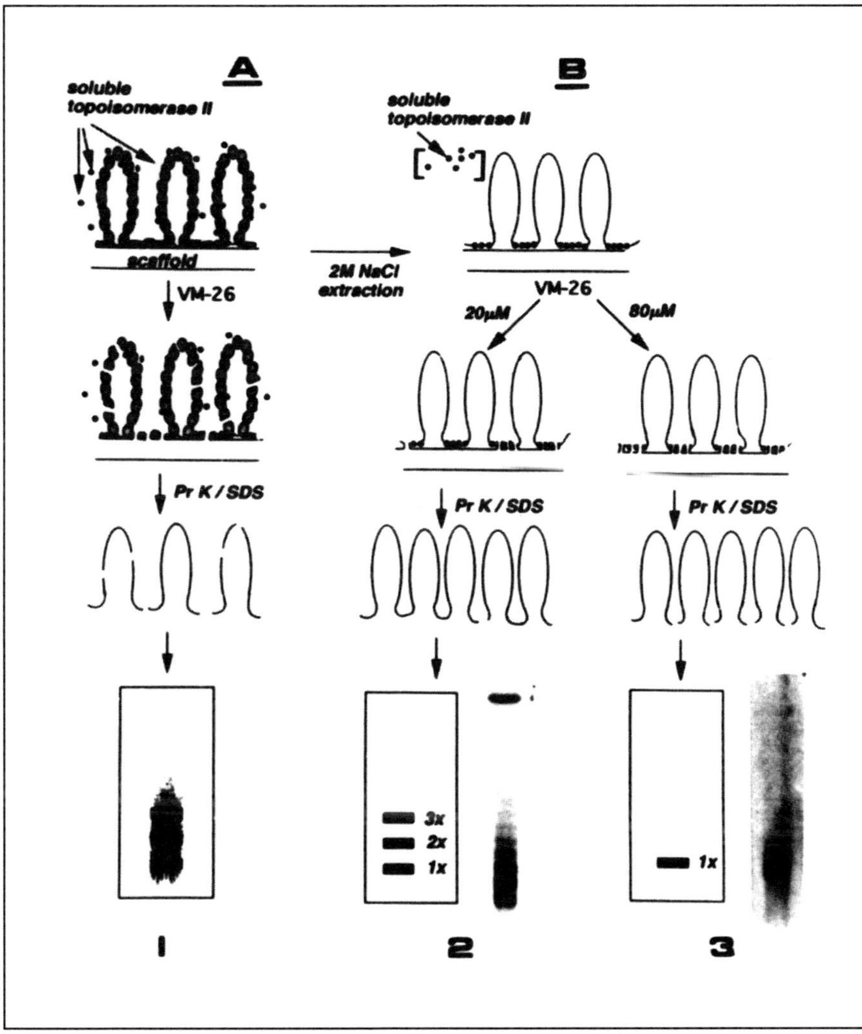

Fig. 5.3. A scheme illustrating the possibility of genomic DNA loops excision by topoisomerase II-mediated DNA cleavage at matrix attachment sites. (A) In living cells, the 30 nm chromatin fibril is arranged into loops by periodic attachment to the nuclear scaffold. The latter contains topoisomerase II which can cut DNA at loop basements. In addition, soluble topoisomerase II can interact with DNA at a number of accessible sites creating an apparently random cleavage pattern (as far as the low-resolution analysis by PFGE is considered). The resulting cleavage pattern (panel 1) will represent the superimposing of this random cleavage pattern over the regular pattern of cleavages at loop basements. (B) With the soluble topoisomerase II removed by high-salt extraction the cleavages can occur only within matrix attachment sites. Varying concentration of VM26 it is possible to cut DNA at all loop basements (panel 3) or less frequently so that multimers of loops are also released (panel 2). (Reprinted with permission of ICSU Press from Razin SV, Gromova II. BioEssays 1995; 17:443-450. © 1995 ICSU Press.)

studied by in situ hybridization of fluorescent probes to the metaphase chromosome halos. Like the experiments on topoisomerase II-mediated DNA loop excision, the in situ hybridization experiments also demonstrated the region of the rDNA nontranscribed spacer located upstream to the transcription start to be preferentially bound to the chromosomal axis or scaffold.[30]

The most important conclusion following from the experiments with topoisomerase II-mediated excision of DNA loops containing nucleolar genes is that the experimental protocol based on a particular model of topological organization of DNA within an eukaryotic cell (namely on the radial loop model of DNA packaging in a eukaryotic chromosome[31-33]) has permitted us to obtain exactly those results that were predicted by the model. Clearly, this not only justifies the suggested experimental approach but proves as well the validity of the initial model.

It should be stressed that the interpretation of the results of the experiments on topoisomerase II-mediated DNA loop excision (Fig. 5.3) does not depend on a particular function which topoisomerase II performs in the nuclear matrix. Several proteins that are likely to interact with DNA at the matrix attachment sites have been identified recently. We shall describe these proteins in chapter 6. Topoisomerase II may be an integral part of the nuclear matrix (chromosomal scaffold) as have been suggested previously[1-3,34,35] or it may be just one of a number of proteins interacting with the ends of DNA loops. Further investigations will perhaps clarify this question. Our data only show that both in living cells and in high-salt-extracted nuclei topoisomerase II interacts with DNA at the loop anchorage sites or, at least, is located in close proximity to DNA at these sites. Otherwise the excision of DNA loops by topoisomerase II-mediated DNA cleavage would not be possible.

5.3. MAPPING OF DNA LOOP ORGANIZATION IN NONNUCLEOLAR REGIONS OF EUKARYOTIC CHROMOSOMES

One specific problem concerning any observation made in experiments with the ribosomal gene cluster is that this observation does not necessarily reflect the situation in the whole genome. Indeed, it is well known that ribosomal genes occupy a special compartment in eukaryotic cell nuclei (nucleoli). One can argue that the spatial organization of nonnucleolar genes may differ from that of nucleolar genes. In order to find out whether this is the case, we have studied the DNA loop organization in an amplified human *c-myc* gene locus[36] and in a 500 kb-long nonamplified region of the *Drosophila* X-chromosome.[37] In these experiments, the separation of the excised loops of chromosomal DNA was combined with indirect end-labeling of large fragments of individual loops obtained after additional cleavage of DNA by rear-cutting restriction nucleases. This permitted us to determine the approximate positions of the loop ends relative to the sites of DNA cleavage by rear-cutting restriction enzymes. In a result, the borders of DNA loops were mapped at the level of PFGE-resolution. As an example, the results of one of the experiments on mapping DNA loop anchorage regions in a segment of the *Drosophila* X chromosome are shown in Figure 5.4. One can see that hybridization of probes complementary to the ends of large Not I fragments of genomic DNA in many cases revealed extra bands (in addition to full size genomic fragments) in samples from cells subjected to high-salt extraction followed by the incubation in a topoisomerase II cleavage buffer supplemented with VM-26. None of the above extra bands was observed in control experiments when VM-26 was not present in the incubation mixture. Obviously, the extra bands originated as a result of DNA cleavage by high-salt-insoluble topoisomerase II of the nuclear matrix. Knowing roughly the sizes of extra-bands one can easily determine the distances between the topoisomerase II cleavage/DNA loop anchorage sites and the Not I site delimiting the end of a genomic fragment recognized by a chosen probe (Fig. 5.4).

In experiments with euchromatic regions of human and *Drosophila* genomes, the preferential targets for topoisomerase II-mediated DNA cleavage generating specific long-range DNA fragmentation again were found to be the same in living cells and in high-salt-extracted nuclei.[36] Furthermore, the choice of a drug (VM-26, VP-16, m-AMSA) to be used for entrapment of topoisomerase II cleavage complexes did not influence the results of the mapping experiments[36] (Fig. 5.5). The latter observation may happen to be of some practical importance because it indicates that the nuclear matrix topoisomerase II constitutes a primary target for a number of presently used antitumor drugs. Thus, our data support the previous conclusion of Fernandes and collaborators[20,21]

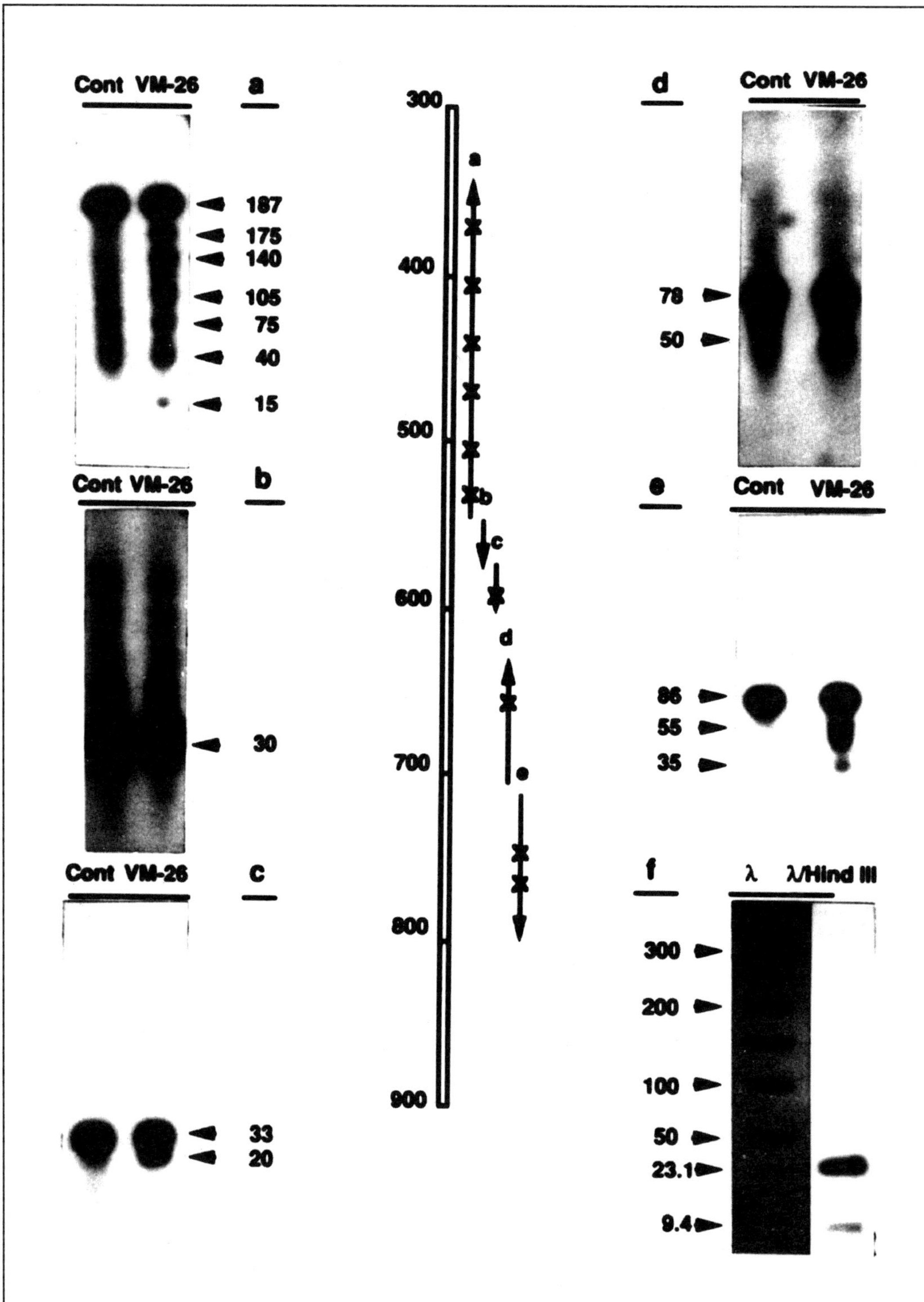
Cont VM-26
a
187
175
140
105
75
40
15
Cont VM-26
b
30
Cont VM-26
c
33
20
300
400
500
600
700
800
900
a
b
c
d
e
d
Cont VM-26
78
50
e
Cont VM-26
86
55
35
f
λ λ/Hind III
300
200
100
50
23.1
9.4

Fig. 5.4 (opposite). Mapping of topoisomerase II cleavage sites within a 500-kb region of Drosophila melanogaster *X-chromosome by indirect end-labeling: Southern analysis of Not I digests of DNA from high-salt-extracted nuclei incubated without (Cont) or with VM-26. Panels a-e show the results of hybridization with probes recognizing DNA located at map positions 543, 550, 580, 705, 715, respectively. Separation of length markers (phage λ DNA concatemers and Hind III fragments of phage λ DNA) is shown in panel f. The full-length fragments and new bands produced by topoisomerase II-mediated cleavage are marked by arrows; the estimated size of each fragment (in kb) is indicated beside the arrows. The map in the center represents the area under study with the positions of full-length restriction fragments (arrows) and of cleavage sites (crosses) identified in each hybridization experiment (a to e). (Reprinted with permission of American Society for Microbiology from Iarovaia O, Hancock R, Lagarkova M et al. Mol Cell Biol 1996; 16:302-308. © 1996 American Society for Microbiology.)*

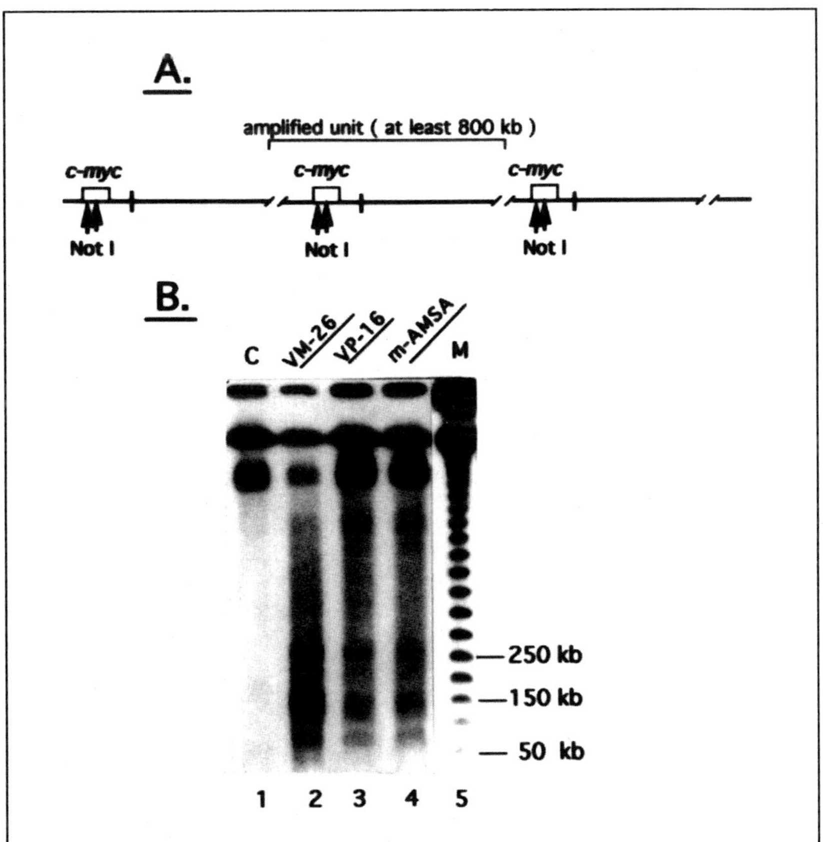

Fig. 5.5. Long-range cleavage of the c-myc amplicon in OC-NYH-VM cells upon inhibition of topoisomerase II by different antitumor agents. (A) Map of the organization of the c-myc amplicon in OC-NYH-VM cells. Position of the c-myc gene is shown by an open rectangle. Arrows show positions of two Not I sites within the c-myc gene. (B) Visualization of cleavage products by hybridization with the third exon of the c-myc gene. Topoisomerase II-mediated DNA cleavage at matrix attachment sites was stimulated by either 40 μM VM-26 (lane 2), 40 μM VP-16 (lane 3), or 40 μM m-AMSA (lane 4). Before electrophoresis all DNA samples were additionally cleaved by Not I restriction endonuclease to permit indirect end-labeling. After separation by PFGE and Southern transfer, the cleavage products containing the third exon of the c-myc gene were identified by hybridization. Line 1 was loaded with material from cells which were not treated with any of the topoisomerase II inhibitors. Line 5 show separation of the phage λ DNA concatemers. The two bands close to the start of the gel are a compression artefact and the full-sized Not I fragment recognized by the chosen probe. The main topoisomerase II cleavage products have sizes of 80, 150 and 260 kb. (Reprinted with permission of The National Academy of Sciences of the United States of America from Gromova II, Thomsen B, Razin SV. Proc Natl Acad Sci USA 1995; 92:102-106. © 1995 by The National Academy of Sciences of the United States of America.)

derived from the results of completely different experiments. It has been reported recently that the utilization of topoisomerase II-specific drugs as chemotherapeutic agents may cause chromosomal translocations resulting in secondary leukemias (reviewed by Albain et al[38] and by Rowley[39]). The results of experiments on topoisomerase II-mediated DNA loop excision suggest that these rearrangements may be a consequence of an illegitimate recombination at the loop basements mediated by topoisomerase II. We shall return to this question in chapter 11.

One of the loop basements mapped in the *c-myc* gene locus roughly colocalized with the *c-myc* gene itself. A high-resolution analysis of the loop anchorage region in the vicinity of the *c-myc* gene made it possible to identify multiple sites of DNA contact with the nuclear matrix topoisomerase II scattered along a 5 kb area including the first two exons of the *c-myc* gene and the upstream sequences (Fig. 5.6).[36] Hence, again a relatively extended loop anchorage region was detected. Although cleavages occurred within the same area in living cells and in high-salt-extracted nuclei, the exact positions of the cleavage sites were changed as a result of high-salt extraction indicating that this extraction may cause a local redistribution of the sites of DNA contact with high-salt-insoluble topoisomerases II within the loop anchorage region[36]

Fig. 5.6 (opposite). High-resolution mapping of cleavage sites generated within the c-myc *gene by topoisomerase II inhibited by different drugs. (A) Map of the* c-myc *gene. Open rectangles denote exon positions. Recognition sites for restriction endonucleases Sfi I, HindIII, Cla I,* Not *I, and* EcoR*I are shown by arrows. Positions of probe 1 (Cla I-EcoRI fragment) and probe 2 (HindIII-Cla I fragment) are indicated by solid bars below the map. The topoisomerase II cleavage region is shown by a hatched rectangle above the map. Horizontal arrow shows the direction of the* c-myc *transcription. (B) DNA samples from cells treated with either 40 μM VM-26 (lane 3), 40 μM VP-16 (lane 4), or 40 μM* m-AMSA *(lane 5) or from control cells not exposed to antitumor drugs (lane 2) were digested with restriction endonucleases Not I and Sfi I. After separation under conditions that allow resolution of DNA sizes from 5-50 kb and Southern blotting, the filter was hybridized with probe 2. Lane 1 contains a 5-kb DNA ladder as size markers; positions of markers are shown at left. (C) Same as for B, but the DNA samples were digested with EcoRI restriction endonuclease, separated by conventional gel electrophoresis and hybridized with probe 1. Phage λ DNA digested with BstEII restriction endonuclease was used as size markers in lane 1. (D) DNA samples were digested with Cla I and HindIII restriction endonucleases and subsequently hybridized with probe I. DNA samples from cells treated with either 40 μM VM-26, 40 μM VP-16 or 40 μM* m-AMSA *are shown in lanes 2-4, respectively. DNA from nontreated cells is in lane 1. (Reprinted with permission of The National Academy of Sciences of the United States of America from Gromova II, Thomsen B, Razin SV. Proc Natl Acad Sci USA 1995; 92:102-106. © 1995 by The National Academy of Sciences of the United States of America).*

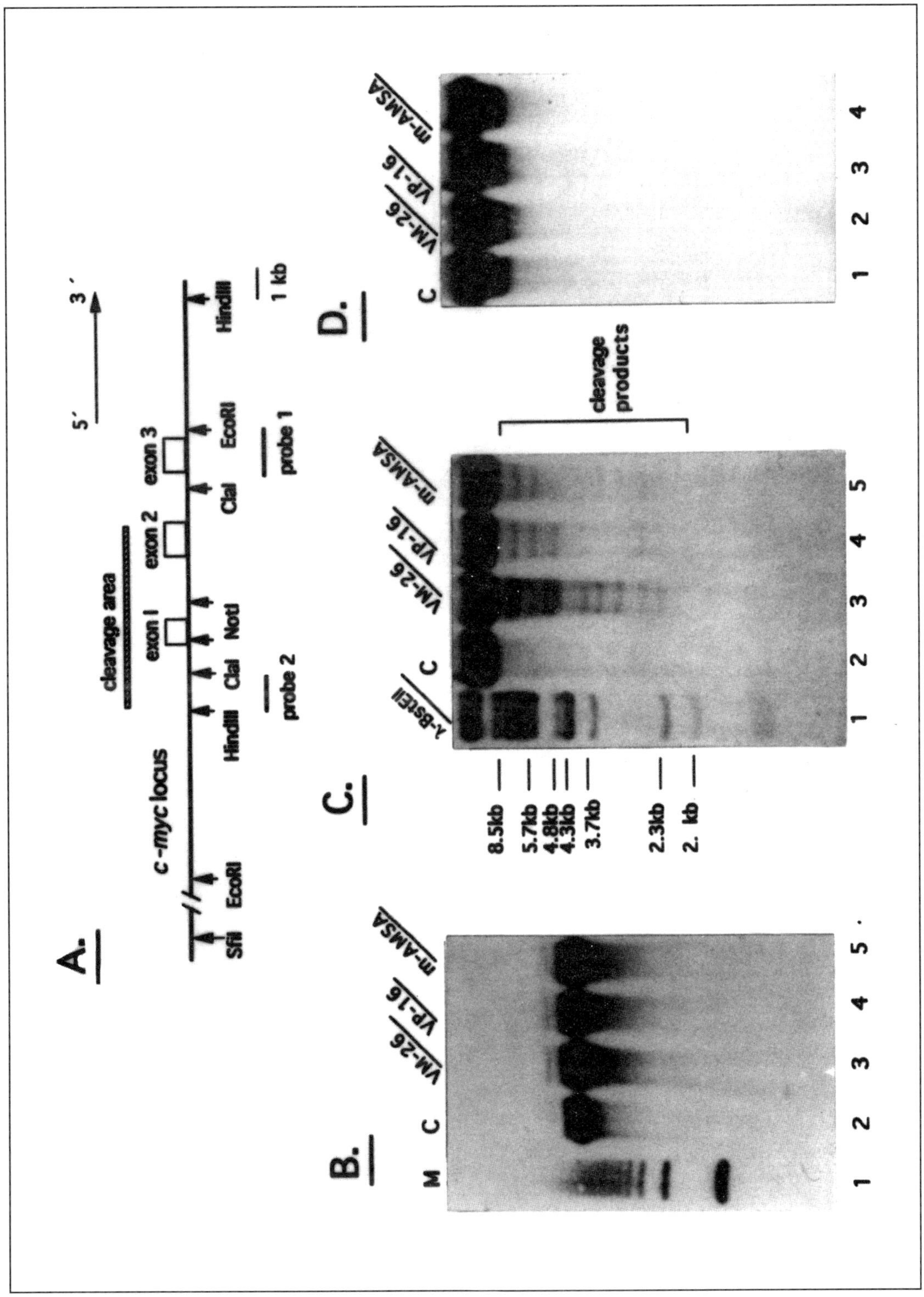
A.
c-myc locus
SfiI
EcoRI
HindIII
ClaI
NotI
exon I
exon 2
exon 3
cleavage area
probe 2
probe 1
1 kb
5′
3′
B.
M
C
VM-26
VP-16
m-AMSA
C.
λ-BstEII
8.5kb
5.7kb
4.8kb
4.3kb
3.7kb
2.3kb
2. kb
cleavage products
D.

Mapping of the DNA loop organization in a previously cloned[37] 500 kb long area of the *Drosophila* first chromosome permitted the identification of eleven loop anchorage regions (LARs) delimiting ten DNA loops ranging in size from 20-90 kb.[37] As a matter of fact, the first loop-domain map of a long area of an eukaryotic genome was constructed. Topoisomerase II-mediated DNA loop excision can be used for construction of similar maps reflecting the organization into loop domains of long areas and eventually, whole chromosomes of the human genome. Being based on the internal principle of genome organization (in contrast to the presently used maps based on a casual distribution of recognition sequences for rear-cutting restriction enzymes), the domain maps may constitute an important tool for subsequent functional studies.

It is of interest that one of the LARs mapped in the *Drosophila* X chromosome was located within an active transcription unit.[37,40] Hence, the transcription machinery can pass somehow the loop anchorage sites (which are most likely colocalized with permanent sites of DNA attachment to the nuclear matrix[29]).

It should be stressed that the large-scale distribution of topoisomerase II cleavage sites on DNA in living cells has long been known to be nonrandom. For example, the *c-myc* gene was reported to be 20 times more susceptible to topoisomerase II-mediated cleavage in vivo as compared to the whole genomic DNA.[41] The results described in this chapter suggest that drastic variations in the efficiency of cleavage of different genes by topoisomerase II may reflect the pattern of DNA interaction with the nuclear matrix. In other words, susceptible genes are likely to be located within matrix attachment/DNA loop anchorage regions while relatively resistant genes may be located in loop DNA.

5.4. CONCLUSIONS

A procedure for mapping DNA loop anchorage regions (LARs) has been developed which permits one to identify high-salt resistant anchorage regions preexisting in living cells. The procedure may be used for constructing long-range maps of the domain-loop organization of large genomic areas, and, eventually of whole chromosomes. Based on the internal principle of genome organization into structural units (loops), these maps may appear to be more useful for functional studies as compared to the presently existing long-range genomic maps based on a casual distribution of recognition sequences for rear-cutting restriction enzymes.

Related Review Articles

Anderson HJ, Roberge M. DNA topoisomerase II: a review of its involvement in chromosome structure, DNA replication, transcription and mitosis. Cell Biol Int Rep 1992; 16:717-724.

Osheroff N. Biochemical basis for the interactions of type I and type II topoisomerases with DNA. Pharmacol Ther 1989; 41: 223-241.

Paoletti C. The localization of topoisomerase II cleavage sites on DNA in the presence of antitumor drugs. Pharmacol Ther 1993; 60:381-387.

Poljak L, Kas E. Resolving the role of topoisomerase II in chromatin structure and function. Trends Cell Biol 1995; 5:348-354.

Razin SV, Hancock R, Iarovaia O et al. Structural-functional organization of chromosomal DNA domains. Cold Spring Harb Symp Quant Biol 1993; 58:25-35.

Razin SV, Gromova II. The channels model of the nuclear matrix structure. Bioessays 1995; 17:443-450.

Razin SV, Gromova II Iarovaia OV. Specificity and functional significance of DNA interaction with the nuclear matrix: New approaches to clarify the old questions. Int Rev Cytol 1995; 162B:405-448.

References

1. Berrious M, Osheroff N, Fisher P. In situ localization of DNA topoisomerase II, a major polypeptide of *Drosophila* nuclear matrix. Proc Natl Acad Sci USA 1985; 82: 4142-4146.
2. Earnshaw WC, Halligan B, Cooke CA et al. Topoisomerase II is a structural component of mitotic chromosome scaffolds. J Cell Biol 1985; 100:1706-1715.
3. Earnshaw WC, Heck MMS. Localization of topoisomerase II in mitotic chromosomes. J Cell Biol 1985; 100:1716-1725.
4. Gasser SM, Laroche T, Falquet J et al. metaphase chromosome structure. Involvement of topoisomerase II. J Mol Biol 1986; 188:613-629.
5. Liu LF, Yang TC, Yang L et al. Cleavage of DNA by mammalian DNA topoisomerase II. J Biol Chem 1985; 256:4805-4809.
6. Wang JC. DNA topoisomerases. Annu Rev Biochem 1985; 60:513-552.
7. Osheroff N, Zechiedrich EL, Gale KC. Catalytic function of DNA topoisomerase II. Bioessays 1991; 13:269-273.
8. Chen GL, Yang L, Rowe TC et al. Nonintercalative antitumor drugs interfere with the breakage-reunion reaction of mammalian topoisomerase II. J Biol Chem 1984; 259:13560-13566.

9. Osheroff N. Effect of antineoplastic agents on the DNA cleavage/religation reaction of eukaryotic topoisomerase II: inhibition of DNA religation by etoposide. Biochemistry 1989; 28:6157-6160.
10. Robinson MJ, Osheroff N. Stabilization of the topoisomerase II-DNA cleavage complex by antineoplastic drugs: Inhibition of the enzyme-mediated DNA religation by 4'-(9-acridinylamino)-methanesulfon-*m*-anisidide. Biochemistry 1990; 29:2511-2515.
11. Udvardy A, Schedl P, Sauder M et al. Novel partitioning of DNA cleavage sites for *Drosophila* topoisomerase II. Cell 1985; 40: 933-941.
12. Sander M, Hsieh TS. *Drosophila* topoisomerase II double-strand DNA cleavage: analysis of DNA sequence homology at the cleavage site. Nucl Acids Res 1985; 13:1057-1072.
13. Spitzner JR, Muller MT. A consensus sequence for cleavage by vertebrate DNA topoisomerase II. Nucl Acids Res 1988; 16: 5533-5556.
14. Fosse P, Paoletti C, Saucier J-M. Pattern of recognition of DNA by mammalian DNA topoisomerase II. Biochem Biophys Res Commun 1988; 151:1233-1240.
15. Rowe TC, Wang JC, Liu LF. In vivo localization of DNA topoisomerase II cleavage sites on *Drosophila* heat shock chromatin. Mol Cell Biol 1986; 6:985-992.
16. Reitman M, Felsenfeld G. Developmental regulation of topoisomerase II sites and DNase I-hypersensitive sites in the chicken β-globin locus. Mol Cell Biol 1990; 10:2779-2786.
17. Udvardy A, Schedl P. Chromatin structure, not DNA sequence specificity, is the primary determinant of topoisomerase II sites of action in vivo. Mol Cell Biol 1991; 11:4973-4984.
18. Drake FH, Hormann GA, Bartus HF et al. Biochemical and pharmacological properties of p170 and p180 forms of topoisomerase II. Biochemistry 1989; 28:8154-8160.
19. Chung TDY, Drake FH, Tan KB et al. Characterization and immunological identification of cDNA clones encoding two human DNA topoisomerase II isozymes. Proc Natl Acad Sci USA 1989; 86:9431-9435.
20. Fernandes DJ, Danks MK, Beck WT. Decreased nuclear matrix DNA topoisomerase II in human leukemia cells resistant to VM-26 and m-AMSA. Biochemistry 1990; 29:4235-4241.
21. Fernandes DJ, Catapano CV. Nuclear matrix targets for anticancer agents. Cancer Cells 1991, 3:134-140
22. Kaufmann SH, Shapper JH. Association of topoisomerase II with the hepatoma cell nuclear matrix: the role of intermolecular disulfide bond formation. Exp Cell Res 1991; 192:511-523.
23. Danks MK, Qiu J, Catapano CV et al. Subcellular distribution of the α and β topoisomerase II-DNA complexes stabilized by VM-26. Biochem Pharmacol 1994; 48:1785-1795.

24. Zini N, Santi S, Ognibene A et al. Discrete localization of different DNA topoisomerases in HeLa and K562 cell nuclei and subnuclear fractions. Exp Cell Res 1994; 210:336-348.
25. Vassetzky YS, Razin SV, Georgiev GP. DNA fragments which specifically bind to isolated nuclear matrix in vitro interact with matrix-associated DNA topoisomerase II. Biochem Biophys Res Commun 1989; 159:1263-1268.
26. Cook PR, Brazell IA, Jost E. Characterization of nuclear structures containing superhelical DNA. J Cell Sci 1976; 22:303-324.
27. Razin SV, Hancock R, Iarovaia O et al. Structural-functional organization of chromosomal DNA domains. Cold Spring Harb Symp Quant Biol 1993; 58:25-35.
28. Iarovaia OV, Razin SV, Hancock R. Domain organization of mammalian ribosomal gene cluster. Dokl. Acad. Nauk (Rus) 1993; 330:120-122.
29. Razin SV, Petrov P, Hancock R. Precise localization of the α-globin gene cluster within one of 20 to 300 kb fragments released by cleavage of chicken chromosomal DNA at topoisomerase II sites in vivo: evidence that the fragments are DNA loops or domains. Proc Natl Acad Sci USA 1991; 88:8515-8519.
30. Bickmore WA, Oghene K. Visualizing the spatial relationships between defined DNA sequences and the axial region of extracted metaphase chromosomes. Cell 1996; 84:95-104.
31. Marsden MP, Laemmli UK. Metaphase chromosome structure: evidence for a radial loop model. Cell 1979; 17:849-858.
32. Getzenberg RH, Pienta KJ, Ward WS et al. Nuclear structure and the three-dimensional organization of DNA. J Cell Biochem 1991; 47:289-299.
33. Pienta KJ, Getzenberg RH, Coffey DS. Cell structure and DNA organization. Crit Rev Eukar Gene Expres 1991; 1:355-385.
34. Boy de la Tour E, Laemmli UK. The metaphase scaffold is helically folded. Sister chromatids have predominantly opposite helical handedness. Cell 1988; 55:937-944.
35. Adachi Y, Luke M, Laemmli UK. Chromosome assembly in vitro: Topoisomerase II is required for condensation. Cell 1991; 64:137-148.
36. Gromova II, Thomsen B, Razin SV. Different topoisomerase II antitumour drugs direct similar specific long-range fragmentation of an amplified *c-myc* gene locus in living cells and in high-salt-extracted nuclei. Proc Natl Acad Sci USA 1995; 92:102-106.
37. Iarovaia OV, Hancock R, Lagarkova MA et al. Mapping of genomic DNA loop organization in a 500-kilobase region of the *Drosophila* X chromosome using the topoisomerase II-mediated DNA loop excision protocol. Mol Cell Biol 1996; 16:302-308.
38. Albain KS, Le Beau MM, Ullirsch R et al. Implication of prior treatment with drug combination including inhibitors of topo-

isomerase II in therapy-related monocytic leukemia with 9;11 translocation. Genes Chromosom Cancer 1990; 2:53-58.
39. Rowley JD. Rearrangements involving chromosome band 11Q23 in acute leukaemia. Semin Cancer Biol 1993; 4:377-385.
40. Surdej P, Got C, Rosset R et al. Supragenic loop organization: mapping in *Drosophila* embryos, of scaffold-associated regions on a 800 kilobase continuum cloned from the 14B-15B first chromosome region. Nucl Acids Res 1990; 18:3713-3722.
41. Riou JF, Lefevre D, Riou G. Stimulation of the topoisomerase II induced DNA cleavage sites in the c-myc protooncogene by antitumor drugs is associated with gene expression. Biochemistry 1989; 28:9104-9110.

CHAPTER 6

Organization of Loop Anchorage Regions (LARs)

From the previous chapter it is clear that relatively long pieces of DNA are in contact with nuclear matrix proteins at the bases of chromosomal DNA loops. In this chapter, we shall analyze the existing data about the DNA sequences located in these regions and about nuclear matrix proteins interacting with these sequences.

6.1. PROPERTIES OF DNA FRAGMENTS PROTECTED FROM NUCLEASES BY THE INTERACTION WITH THE NUCLEAR MATRIX PROTEINS

Attempts to identify the consensus DNA sequence essential for the anchorage to the nuclear matrix can be traced back to the late seventies. Initially, the renaturation properties of relatively short DNA fragments isolated from nuclear matrices or scaffolds of metaphase chromosomes were studied. Although in some experiments an enrichment of these DNA fragments in repetitive sequences was observed,[1-6] contradictory results were reported by other authors.[7,8] The most discouraging fact was, however, that even in those cases when the sequence organization of nuclear matrix DNA was found to be nonrandom, no particular DNA sequence element responsible for the attachment to the nuclear matrix was identified.[4,9,10] This failure may have several reasons. One of them is that such an element could not be identified by the analytical approaches used, namely, reassociation kinetics analysis and cross-hybridization experiments. Indeed, conserved recognition sites for sequence-specific DNA-binding proteins in most cases are very short

(4-8 bp). If a similar recognition site or, more likely, a combination of such recognition sites is necessary for matrix binding, they cannot be identified by cross-hybridization. Furthermore, it may well be that a matrix attachment element (if any) is not a distinct DNA sequence, but rather a kind of sequence motif, for example, a DNA sequence able to form (under certain conditions) a noncanonical secondary structure. To check the above mentioned possibilities, it seemed reasonable to clone randomly chosen nuclear matrix DNA fragments and to compare their nucleotide sequences. By the time this idea was given serious consideration, it had been found that the interaction with the nuclear matrix protected from nucleases only rather short DNA fragments ranging in size from 50-200 bp. These fragments could not be removed from matrices by repeated extractions with 2M NaCl solution or even by density centrifugation in a CsCl gradient.[9] However, it was not clear whether these fragments originated from permanent or from function-dependent attachment sites. In order to simplify the interpretation of the results, we have cloned and sequenced nuclear matrix DNA fragments from chicken erythrocytes (i.e., fragments released from permanent attachment sites).[11,12] In other laboratories, short nuclear matrix DNA fragments isolated from cells synchronized at the border of G1/S phases were cloned and sequenced.[13-16] Perhaps surprisingly, in all cases a significant portion or even a majority of short nuclear matrix DNA fragments were found to possess a degenerated sequence organization, as followed from the presence of short perfect and imperfect repeats and simple motifs. This degenerated sequence organization may be easily recognized in the dot-matrix maps of internal homologies. As an example, four such maps reflecting the sequence organization of some chicken erythrocyte nuclear matrix DNA fragments are shown in Figure 6.1. No genes (or parts of genes) were apparently present in the preparations of short nuclear matrix DNA fragments even when these fragments were isolated from active nuclei. This may indicate that the association of transcriptionally active DNA with the nuclear matrix is resistant to high-salt extraction only when relatively long fragments of DNA interact with the matrix at multiple sites. Indeed, in experiments demonstrating an association of transcriptionally active genes with the nuclear matrix[17-22] the restriction nucleases were used for cleavage of DNA from the nucleoid core.

CO 326 245 b.p. A+T - 33%

GACCGCTTCG GGTGTCCTTC GGGTCCGTTC GGCGGCGTTC GGGTCACTGC AGATCACTTC
GGGTGTCCTC GGGTCCGGTT CGGGTCCGTT GGCCGGCGTT CGGGTGACCG CAGAGAGCTT
CGGGTCTCTT CGGGTCCGTT CGGCGGCGTT CGGGTCACTG CAGACCGCTT CGAGTCTTTT
CGGGTCCGGT TCGGGTCCGT TCGGCGGCGT TCGGGTCACT GCAGACAACT TCGGAACTCT
TCGGG

CO 62 220 b.p. A+T - 20%

GGAACGGAAC GGAACGGAAC GGAACGGAGC GGAGCGGAGC AGAGATGCCA CCGCCGCCGT
CGGCCCGCCG TGACGCCGTG CGCCGGCGGG GGCGGCGGGG AGAGCCGGGG GGGGAACCCG
AGTGCCGCAG CTCCGCCGGT ACCGGAACCC CCGGAGAGGG TGCGTGGCAC GGGGGGGTGC
GGGAGGCGAG GGGGTGGCGG GAGGGCACGG CAGCAGGGGG

CO 326 **CO 62**

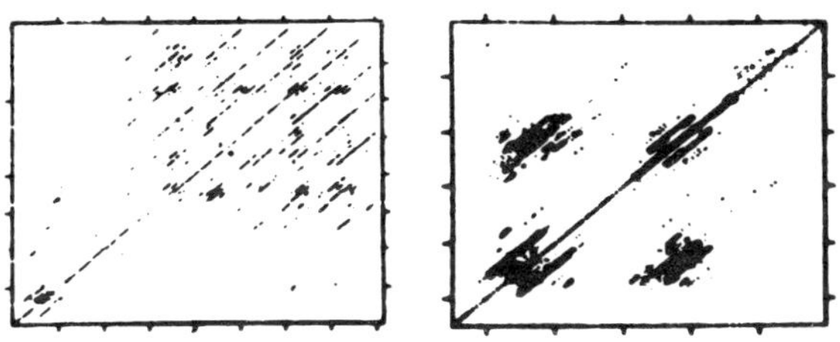

CO 30 160 b.p. A+T - 26%

ATCCGGGGGG GGTCGTGGGG GGGGGGAGTC CCCAATTCCA AGGGGGGGGT CGGAGGGTGG
GGGGGGTCCC CATTCCAAGG GGGGGGCCGG AAGGGGGAGG GTCCCCATTT CTCAGGGGGG
GGTCGGAGGG GGGGGTCCCC ATTCCCAACG GGGGAGGGGG

CO 73 517 b.p. A+T - 42%

CCCCCCAGTG GGGAGGTCCT GTGGGGGGGT TTGGGGGGGT CCTAGGGAGG TTTTGGGGGG
GATAGTGGGG TTTGGGGGGG TCCTATGGGG GATGGGGGGG TTCTGGAGGG GTCCAAGGGG
GGATACTGGG GTTTGGGGGG GTCCTATGGG GAATGGGGGG GTCCCTGAGG GGTCCAAGGG
GGGATAGTGG GGTTTGGGGG GGCTGGGTCC GTTGGGGATG GGTAGGAATG CGGCCCACTG
TGAGGTCACG GGGTGACTGT GACTCTTCGG TGCCATCACG AAGCACACAG ATAGGTGGAG
GTTTGCTGCC CCGTGGCTAT TGCGTCAGTA TTTTGGGCAT TTGCCTGCCA GAGTGCACCC
GAGGACTTTC AGTGGGCCAT TTTCCCAAGC AGTGACAAAT AAAGGATTCT TGAAGCAAGC
AGTAGAGCAG CTTTGGTGTT GCCGTTGATC AGTGCAGTAG CTGTGACCGT GCCGGTATTT
GATAGTAGTG TCAGCAAAGT AAATGCTGTG CTGGGTA

CO 30 **CO 73**

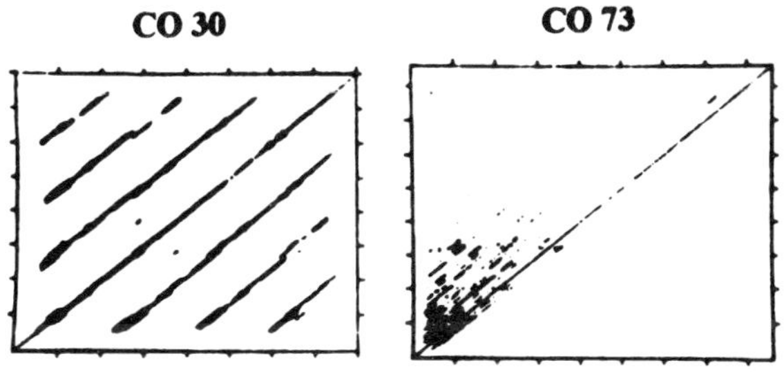

Fig. 6.1. Sequences of several cloned nuclear matrix DNA fragments with the dot-matrice maps of internal homologies. (Reprinted with permission of Academic Press, Inc. from Razin SV, Gromova II Iarovaia OV. Int Rev Cytol 1995; 162B:405-448. © 1995 Academic Press, Inc.)

Consequently, the nuclear matrix DNA fragments were usually several kb long.

6.2. RECONSTITUTION IN VITRO OF COMPLEXES OF DNA WITH THE NUCLEAR MATRIX OR WITH THE NUCLEAR MATRIX PROTEINS

It has long been known that a number of nuclear matrix proteins can bind DNA on western blots (see, for example, refs. 23-25). This binding was not, however, specific in respect to the DNA sequences involved, although some preference for AT-rich DNA sequences was observed already in initial experiments.[23] It was logical to suggest that the specificity of DNA binding can be lost as a result of disassembly of multiprotein complexes present in the nuclear matrix. Hence, attempts were made to reconstitute these complexes from purified nuclear matrix proteins and DNA. Although in these experiments an assembly of some DNA-protein complexes was achieved, the complexes were not resistant to the high-salt extraction and no sequence specificity of DNA binding was observed.[26] More interesting observations were made when binding of DNA to an isolated nuclear lamina was studied.[27] It was found that the lamina contained binding sites capable of detaining mainly AT-rich DNA sequences. However, the binding specificity was not properly studied. At the same time, Garrard and collaborators found that a subset of cloned DNA sequences could be detained by some affinity sites on isolated nuclear matrices.[28-30] Binding of DNA to the residual nuclear structures was disturbed upon extraction with concentrated or even medium salt solutions. The DNA sequence elements selectively (in the presence of a nonspecific competitor) detained by isolated nuclear matrices were called Matrix Associated Regions or MARs.[28] Binding of these DNA sequences to the nuclear matrices was not tissue- or even species-specific.[29,31] The number of MAR-binding sites was estimated as ~50,000 per nucleus. Although most, if not all, DNA fragments found in complexes with nuclear matrices were relatively rich in AT-pairs, some other DNA fragments with a similar nucleotide composition were not detained by nuclear matrices.[32] Hence, it was logical to assume that DNA fragments able to interact with the sites of affinity on the nuclear matrix possessed some common property determined at the level of nucleotide sequence. Comparison of nucleotide sequences of several cloned DNA fragments containing MARs did not reveal any prominent homology.[33] Recent evidence suggests that an essential feature of MARs is the ability to melt under relatively mild conditions.[34] This ability is probably determined by the presence of (AT)n motifs[34] and ATATTT motifs.[28] Another characteristic feature of MARs determined by the presence of numerous short homopolymeric runs of A and T is a narrow DNA minor groove.[35] The same short tracks of A and T seem to determine the frequent occurrence of permanently curved (bent) DNA regions within MAR elements.[36]

In experiments by Bode and collaborators a distinct group of MARs with a balanced content of AT and GC base pairs was identified.[37] These MARs contain multiple direct repeats and polypurine tracts. Furthermore, they are rich in CA/TG and AG/CT dinucleotide repeats the presence of which has been reported to correlate with a high potential for the formation of cruciform structures.[37] It is obvious that MARs of this type are in a way similar to the nuclear matrix DNA fragments isolated from high-salt-extracted nuclei extensively treated with nucleases (see section 6.1). It is of interest that MARs with a balanced content of AT and GC base pairs are located in highly expressing chromosomal sites and serve as preferential targets for the integration of retroviruses.[37]

The distribution of MARs in the genome has been intensively studied. MARs were found to colocalize frequently with regulatory sequences (such as enhancers), recombination hot-spots and the borders of DNase I sensitive genomic domains.[28,38-42] In yeast cells, MARs seem to constitute an essential part of autonomously replicating sequences.[31] At the same time, some MARs were mapped within transcriptional units and even within structural genes.[43-46] Although it was assumed initially that all specific DNA-binding sites at the nuclear matrix are more or less equal,[29] more recent data suggest that these binding sites are in fact heterogeneous.[47-49] In particular, there are distinct populations of affinity sites which recognize supercoiled DNA[48,49] and single-stranded DNA.[47,48]

Very interesting observations were made in the laboratory headed by R. Berezney.[47] It was found that initial associations between exogenous DNA and the nuclear matrix, which are sensitive to a salt concentration, could be rendered the high-salt-resistant ones after incubation of complexes at 37 °C. As has been discussed previously (chapter 1), this incubation is one of the approaches commonly used for stabilization of an internal nuclear matrix (for a review see ref. 50). In the same investigation,[47] it was demonstrated that only a part of nuclear matrix DNA binding sites recognized MARs preferentially to other DNA sequences. A major population of affinity sites recognizes preferentially single-stranded DNA molecules. The number of these binding sites was estimated as approximately 150,000 per individual matrix.

In the present literature MARs are also called SARs (scaffold attachment regions). This term was originally suggested by Laemmli and collaborators[51] for designation of DNA fragments that remained associated with the nuclear scaffold isolated by extraction of nuclei with lithium diiodosalicylate (LIS). It was assumed that this extraction extended chromatin loops by disrupting nucleosomes but did not affect the integrity of the loop anchorage sites. Following the strategy used in experiments with high-salt-extracted nuclei,[52-54] one can cut the loop DNA with restriction nucleases and isolate DNA sequences remaining bound to the LIS-extracted nuclear scaffold (Fig. 6.2A). These sequences (called Scaffold Attachment Regions, SARs) turned out to be indistinguishable from MARs identified by in vitro binding of cloned DNA fragments to the isolated nuclear matrix.[28,55] It should be stressed here that it has not been specially checked whether DNA is indeed organized into constrained domains (closed loops) in LIS-extracted nuclei. For example, experiments on titration of LIS-extracted nuclei with ethidium bromide have never been done. Hence, it may as well happen that the treatment of isolated nuclei with LIS in a hypotonic buffer destroys all specific interactions of DNA with the nuclear matrix (scaffold). Even in this case the DNA will coprecipitate with the nuclear matrix (scaffold) because it is difficult to separate the extremely long DNA chain from the matrix

Fig. 6.2 (opposite). Two alternative ways to explain a possibility to isolate specific DNA sequence elements in a complex with the nuclear matrix after extraction of nuclei with LIS. (A) As has been proposed by Mirkovich et al,[50] treatment of nuclei with LIS solution extracts histones and extend DNA loops but does not affect the loop anchorage sites. After treatment of nuclear halos with restriction nucleases, complexes of the nuclear matrix with the anchored DNA fragments are separated from the cleaved-off distal parts of DNA loops by differential centrifugation. (B) Extraction of nuclei with LIS solution destroys preexisting DNA anchorages on the nuclear matrix. However, a long DNA molecule entangled in a nuclear matrix network cannot be easily separated from the latter. After cleavage of DNA with a restriction nuclease, short DNA fragments are eluted from the nuclear matrix. Eventually the fragments bearing MARs interact with MAR-binding sites on the nuclear matrix. These complexes may be separated from free DNA fragments by precipitation.

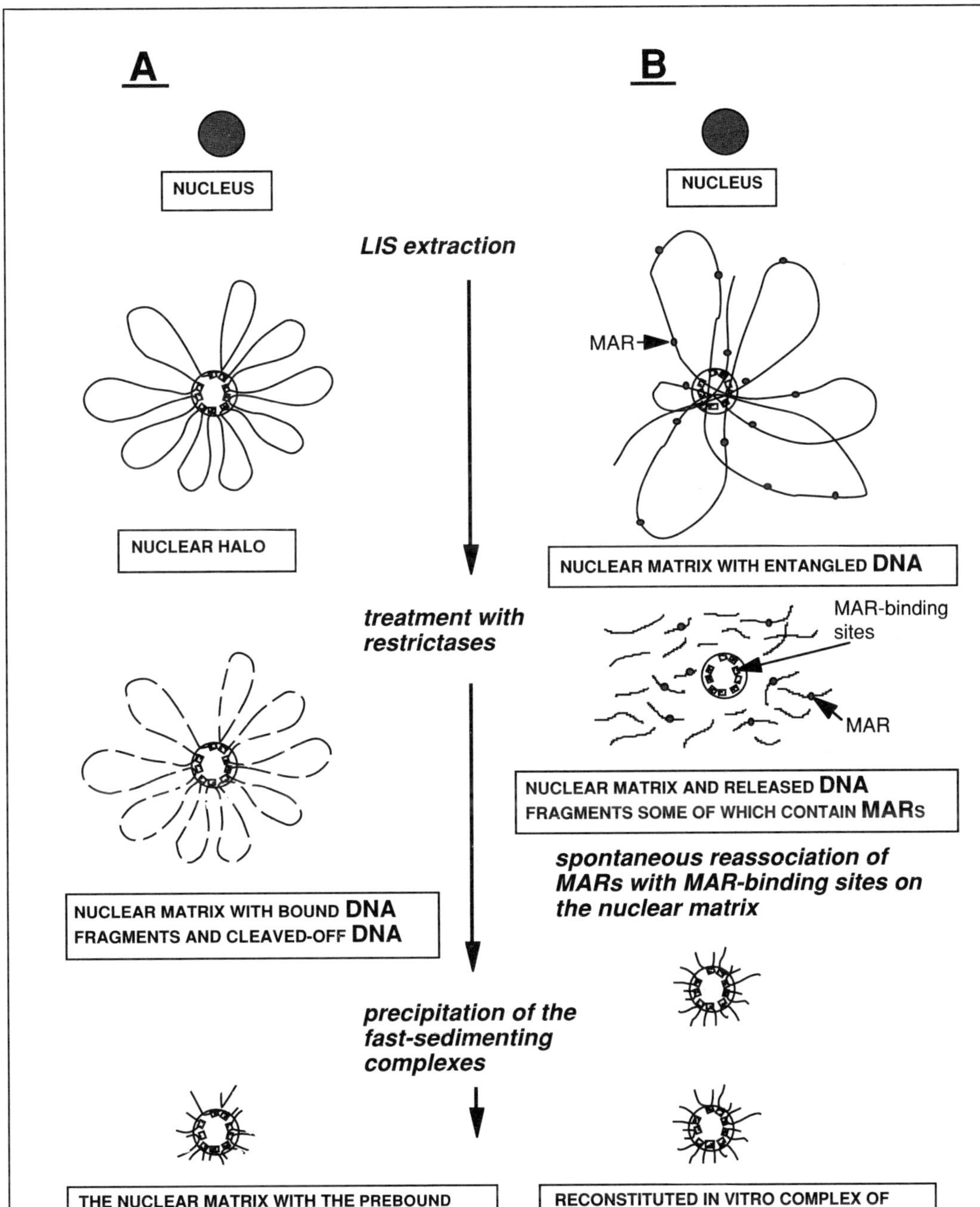
A
B
NUCLEUS
NUCLEUS
LIS extraction
MAR
NUCLEAR HALO
NUCLEAR MATRIX WITH ENTANGLED DNA
treatment with restrictases
MAR-binding sites
MAR
NUCLEAR MATRIX AND RELEASED DNA FRAGMENTS SOME OF WHICH CONTAIN MARs
spontaneous reassociation of MARs with MAR-binding sites on the nuclear matrix
NUCLEAR MATRIX WITH BOUND DNA FRAGMENTS AND CLEAVED-OFF DNA
precipitation of the fast-sedimenting complexes
THE NUCLEAR MATRIX WITH THE PREBOUND DNA FRAGMENTS
RECONSTITUTED IN VITRO COMPLEX OF MARs WITH THE NUCLEAR MATRIX

network. As has been demonstrated previously, the MAR/SAR binding sites are resistant to LIS treatment.[55,56] These binding sites may interact with target DNA sequences during treatment of LIS-extracted nuclei with restriction enzymes (Fig. 6.2B). Incubation of isolated nuclear matrices with a mixture of DNA sequences under study in a buffer suitable for DNA treatment with restriction nucleases is normally used for in vitro binding of MARs to the affinity sites on the nuclear matrix isolated either by high-salt extraction or by the LIS extraction procedure.[28,29,55] Genomic DNA fragments which do not contain SARs can substitute a nonspecific competitor used in the classical[28,29] in vitro-binding assay. Complexes of SARs with the nuclear scaffolds prepared by LIS-extraction are not resistant to reextraction with the same solution. In contrast, DNA fragments isolated in complex with the nuclear matrix after high-salt extraction cannot be washed out by repeated treatments with a high salt solution.[9] Hence, it does not seem likely that complexes of SARs with affinity sites on nuclear matrices/ scaffolds (even if they preexist in living cells) survive the first extraction of nuclei with LIS solution. The LIS extraction procedure may be therefore considered as a technical variant of the in vitro binding assay for identification of DNA sequences capable of interacting with the affinity sites on an isolated nuclear matrix. This conclusion is corroborated by the finding that at physiological ionic strength some SARs can easily be electroeluted from nuclei.[57,58] Thus, complexes of these SARs with the nuclear matrix can hardly preexist in living cells, although these complexes may be identified by the standard LIS extraction procedure.[57,58]

For the present discussion, the most important question is whether at least some MARs/SARs participate (possibly along with other genomic elements) in the organization of DNA loop anchorage sites. Recent experiments of Strick and Laemmli[59] presented functional evidence for the involvement of MARs/SARs in condensation of metaphase chromosomes and maintaining their shape. These authors synthesized a protein containing multiple SAR/MAR-binding sites by amplification of the SAR/MAR-binding domain of HMG-1. This protein termed MATH (for Multi-AT-hook) was found to interfere with the formation of mitotic chromosomes and to cause reshaping of already formed chromatids into spherical bodies. The suppression of chromosome condensation by MATH can be dampened by the addition of an excess of exogenous MARs/SARs. The authors conclude that the interaction of

MARs/SARs with chromosomal proteins (which becomes impossible after the formation of more stable complexes with MATH) is essential both for the formation of mitotic chromosomes and for maintaining their shape.[59]

In the previous chapter the experiments on mapping LARs in a 500 kb region of the *Drosophila* first chromosome[60] have been described. The distribution of MARs/SARs in the same region was studied previously by Miassod and collaborators.[61,62] Comparison of the two sets of data (Fig. 6.3) made it possible to conclude that a portion of MARs/SARs does participate in the organization of loop anchorage sites. Indeed, with a single exception (the loop anchorage site at map position 470), the loop anchorage sites mapped by the topoisomerase II-mediated DNA loop excision protocol colocalized or at least partially overlapped with DNA fragments bearing SARs. The reliability of this correlation is supported by the results of statistical analysis. The probability of obtaining occasionally the observed correlation between the distribution of loop anchorage sites and SARs is as low as 0.0033.[60] There is, however, an excess of SARs which appear to be located in loop DNA. Hence, it is clear that at best a SAR may constitute a part of a loop anchorage site but the presence of a SAR is not sufficient to provide for high-salt-resistant DNA loop anchorage at the nuclear matrix. It may happen that SARs present in loop DNA constitute potential attachment sites. Within these potential attachment sites SARs may be "blocked" by interaction with the soluble MAR/SAR-binding proteins, such as SAF-B (see below). It is significant that no dependence between the "strength" of SARs, as determined by the LIS extraction procedure,[61,62] and the location of these SARs at or close to the actual DNA loop anchorage sites was observed. As a matter of fact, only one strong SAR colocalized with a loop anchorage site determined by the topoisomerase II-mediated DNA loop excision protocol (Fig. 6.3).

6.3. MAR/SAR-BINDING PROTEINS AND OTHER PROTEINS INTERACTING WITH DNA AT MATRIX ATTACHMENT SITES

It was natural to expect that affinity sites on the nuclear matrix capable of recognizing MAR/SAR elements are composed of proteins that are able to recognize the same elements. In an attempt to identify these proteins, the ability of different nuclear matrix proteins immobilized on filters (Western blots) to bind

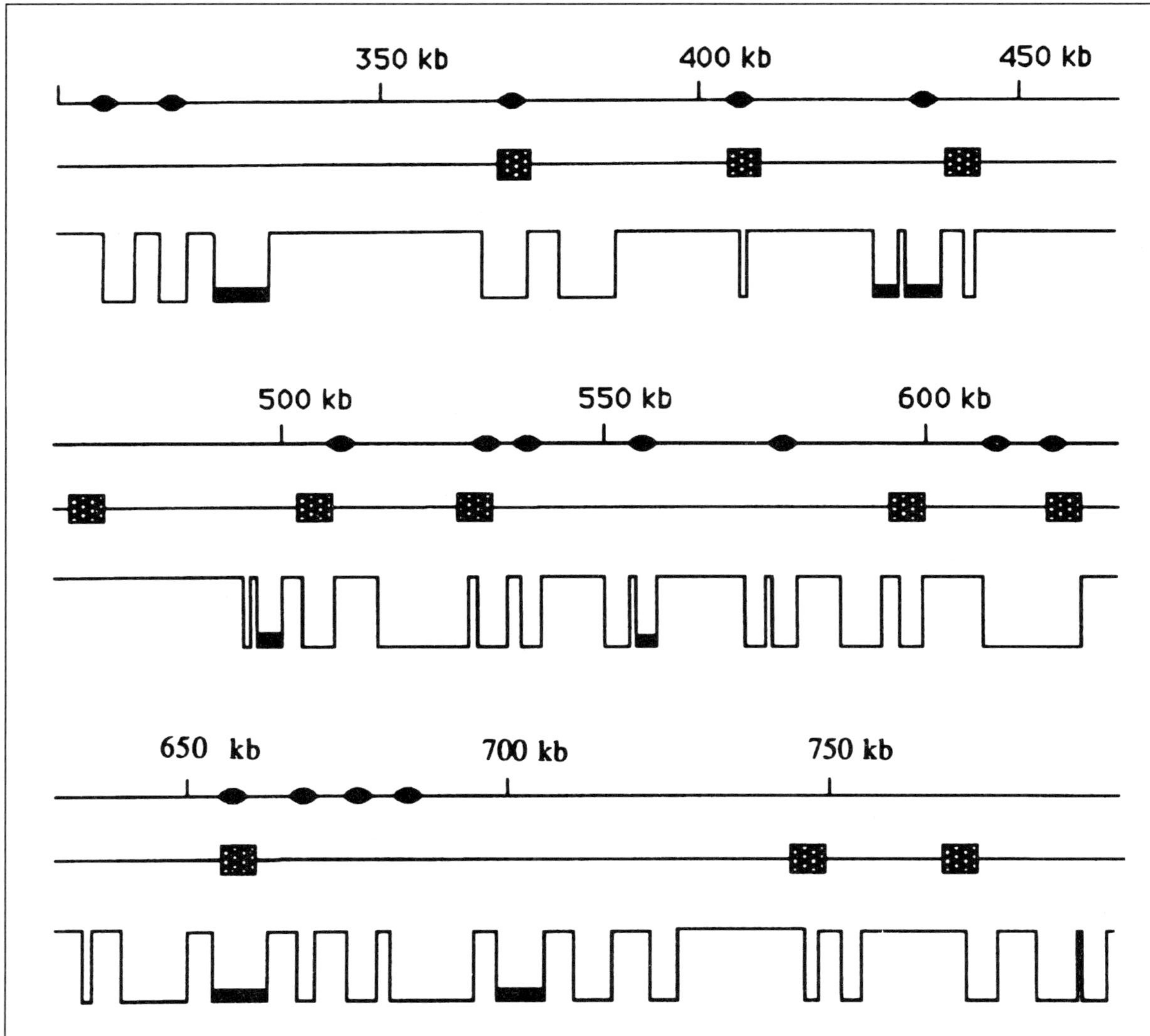

Fig. 6.3. A map of the 500 Kb region of the Drosophila X-chromosome showing the positions of DNA loop anchorage sites mapped by topoisomerase II-mediated DNA cleavage (shaded rectangles on the line below the scale) and the positions of SARs (lower portions of the broken bottom line). The upper and lower parts of the broken lines represent correspondingly the SAR-free regions and regions bearing SARs of different strengths, the strong SARs being shown by thick line. Filled ovals on the scale line show positions of DNA sequences possessing ARS activity in yeast cells. (Reprinted with permission of American Society for Microbiology from Iarovaia O, Hancock R, Lagarkova M et al. Mol Cell Biol 1996; 16:302-308. © 1996 American Society for Microbiology).

preferentially MARs/SARs in the presence of a competitor DNA was tested. Perhaps surprisingly (because MAR-binding sites on the nuclear matrix could be composed of several proteins), this approach has permitted the identification of several specific MAR/SAR-binding proteins which are described below.

Attachment region binding protein

Attachment region binding protein (ARBP), was first identified in nuclear matrices of chicken oviducts by Strätling and collaborators.[63] This 95 kDa protein binds homologous (chicken) and also heterologous (*Drosophila*, mouse, human) MARs in the presence of up to 10000-fold excess of competitor *E. Coli* DNA. Non-MAR DNA fragments with a similar content of AT pairs as in MAR-fragments were not bound by ARBP in a specific fashion. ARBP is an abundant nuclear protein present in about 100,000 copies per oviduct cell nucleus. Fractionation of nuclear matrices into a lamina and an internal matrix according to Kaufmann et al[64,65] demonstrated ARBP to be a component of the internal nuclear matrix.[63] Recent studies have demonstrated that, within different MARs, ARBP preferentially recognizes a sequence motif containing the central GGTGT element flanked by AT-rich sequences. Different substitutions (including single-base substitutions) in the core sequence element resulted in the loss or a significant reduction of ARBP binding.[66]

SAF-A - p120 - SP120 - hnRNP U

Several authors reported the identification of apparently the same MAR-binding protein of 120 kDa in nuclear matrices of different cells.[63,67,68] The protein was purified to homogeneity and found to possess no nuclease activity or activity of topoisomerase I or II.[67] Similarly to ARBP, SAF-A is able to bind MARs in a specific fashion in the presence of a vast excess of the competitor. Furthermore, SAF-A was found to be a component of the internal nuclear matrix. An interesting feature of SAF-A is the ability to form large aggregates which can organize exogenous pure DNA containing MARs into loops (Fig. 6.4). This feature resembles the properties of the yeast repressor-activator protein (RAP-1).[69] This protein, originally isolated from yeast nuclear matrix protein extracts is able to promote the loop formation with DNA bearing

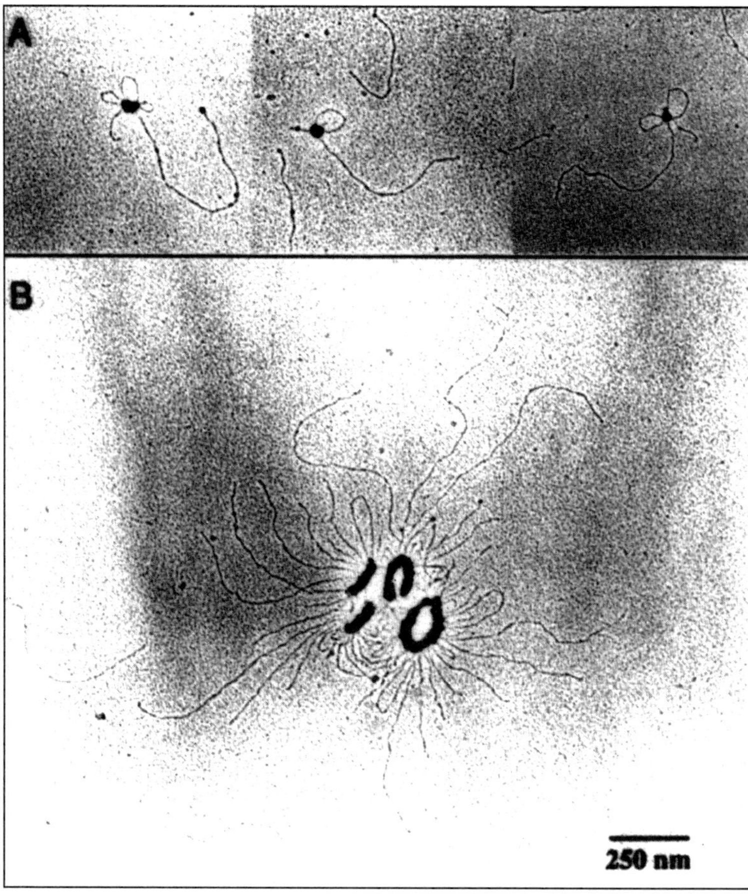

Fig. 6.4. Reconstitution of DNA loops by interaction of SAF-A with an MAR-containing DNA. Electron micrographs of complexes. (A) DNA molecules bound to nonfilamentous SAF-A multimers. (B) DNA molecules bound to filamentous multimers of SAF-A. (Reprinted with permission of IRL Press Ltd. from Romig H, Fackelmayer FO, Renz A et al. EMBO J 1992; 11:3431-3440. © 1992 EMBO/IRL Press Ltd.).

MARs.[69] Most surprisingly, the recent studies have demonstrated that SAF-A is identical to hnRNP U, a protein known to participate in packaging of hnRNP into ribonucleoprotein particles.[70,71] An important thing is that, as shown by UV-crosslinking, SAF-A interacts with DNA in vivo.[70] Hence, this protein is likely to play a dual function in eukaryotic cell nuclei participating in packaging of both hnRNA and DNA.

SAF-B

SAF-B, an abundant MAR-binding nuclear protein, was purified to near homogeneity from HeLa cell nuclear extract.[72] This protein is associated with chromatin but is not a component of the nuclear matrix. The expression of SAF-B was detected in all human tissues tested.[72] The gene encoding SAF-B has been cloned and sequenced. As predicted from the c-DNA sequence, SAF-B contains 849 amino acids, is rich in charged residues, has two putative nuclear localization signals and has no significant homology to any known protein.[72]

Lamins A, B and C, NuMA and desmin

Within this group of MAR-binding proteins we can see a good example of limited suitability of the protein Western blot assay for the identification of DNA binding proteins. Using this method, Van Driel and collaborators demonstrated that lamin B belonged to the group of MAR-binding proteins.[73] However, it was not clear whether other lamins (which did not bind MARs on Western blots) possessed the same properties. Denaturation of a protein in the course of SDS-electrophoresis can alter significantly its properties, including the ability to bind some DNA sequences in a specific fashion. Indeed, when the ability of the polymers of isolated lamins A and C to bind MARs in solution was tested, these polymers were found to possess MAR-binding sites, as well as a polymer of lamin B.[74] Furthermore, it turned out that other proteins structurally related to the intermediate filament proteins (NuMA and Desmin) also possess the property to bind MARs in a specific fashion. The preference for MARs as compared to other AT-rich sequences in all these cases was, however, less pronounced than in the case of the ARBP protein. It seems that the ability of lamins, NuMA and Desmin to form coiled-coil polymers is important for MAR-binding.[74] Recent evidence indicates that the polymerized α-helical rod domain of lamins provides by itself the specific binding site for MARs.[75]

Single-stranded DNA competed with MARs for the binding of lamins, NuMA and Desmin much more efficiently than double-stranded DNA,[74] suggesting that the high potential of MARs for base unpairing[34] is essential for the interaction. On the other hand, interactions of the proteins with the DNA minor groove also seemed to be involved in binding of lamins by MARs, as drugs interacting with the DNA minor groove (distamycin) suppressed this binding.[74]

Special AT-rich sequence binding protein 1

Special AT-rich sequence binding protein 1 (SATB1), a protein with an apparent molecular weight of 103 kDa, was originally found by screening a cDNA expression library from human testis with an artificial MAR sequence characterized by a high base-unpairing potential.[76] Subsequent analysis demonstrated that this protein was predominantly expressed in thymus. A mouse gene encoding a MAR-binding protein sharing 98% homology in the amino acid sequence with SATB 1 was cloned in the same laboratory.[77] SATB1 was found to recognize a specific DNA sequence context rather than any DNA sequence consensus. This context consists of AT-rich DNA sequences in which cytosine residues are found exclusively in one strand and guanine residues in the other (ATC-motifs). Sequences of this kind have a strong tendency to unwind by extensive base unpairing[78] and are frequently found within various MARs. DNA sequences similarly rich in AT-pairs, but lacking the above mentioned asymmetry in the distribution of C and G residues are not preferentially recognized by SATB1. It is of interest that ATC motifs are also recognized by other tissue specific MAR-binding proteins, such as B-cell regulator of IgH transcription ("**Bright**")[79] and p114 MAR-binding protein which is selectively expressed in human breast carcinomas.[80]

Deletional analysis made it possible to map the DNA-binding domain of SATB1 composed of 150 amino acids.[77] Both the N-terminal end and the C-terminal end of this domain were found to be absolutely necessary for DNA binding. A deletion of no more than 21 amino acids from either of the two ends of the DNA-binding domain eliminated almost completely the DNA-binding activity.[77] The functional significance of the internal area is currently not clear. Using a random peptide bacteriophage display library, Kohwi-Shigematsu and coauthors have identified a nonamer peptide (CRQNWGLEGC) capable to recognize with high affinity the ATC-motifs in DNA.[81] Regions sharing homology with the above peptide were found both at N-terminal and C-terminal ends of the SATB 1 DNA-binding domain. Deletion of the N-terminal region of similarity to the CRQNWGLEGC peptide or substitution of this region with a random peptide resulted in a loss of MAR-binding activity. In contrast, substitution of the same region with the CRQNWGLEGC peptide did not affect the MAR-binding activity. Analysis of point mutations has demonstrated that within the CRQNWGLEGC peptide the Arg and the Glu residues are critical for MAR binding.[81]

Nucleolin

Nucleolin is a major nucleolar protein of dividing cells which is likely to regulate the transcription of ribosomal genes and the assembly of ribosomes. Kohwi-Shigematsu and collaborators have demonstrated that nucleolin is also a MAR-binding protein. This 100 kDa protein was isolated from extracts of human K562 cells by affinity chromatography on a column with an immobilized synthetic MAR containing the AATATATTT motif.[82] Nucleolin efficiently binds MAR sequences with a high base unpairing potential. However, it does not recognize mutated versions of the same MARs which do not possess a high base-unpairing potential. Nucleolin also binds with high affinity single-stranded DNA (with preference for the T-rich strand as compared to the A-rich strand) and also RNA. This protein is not an exclusive component of the nuclear matrix. It is partitioned between the high-salt-soluble and high-salt-insoluble nuclear compartments.[82]

DNA Topoisomerase II

DNA Topoisomerase II, one of the major proteins of both nuclear matrix and chromosomal scaffold, was also demonstrated to be able to form complexes with MAR elements. The interaction of this enzyme with MARs seems to have a cooperative fashion and depends on the accessibility of the DNA minor groove.[83] Although the sequence specificity of DNA binding by topoisomerase II is relatively low, it may be significantly increased by the formation of a complex between topoisomerase II, scaffold protein II (SC II) and another presently uncharacterized protein of 48 kDa.[84] The functional significance of the above complex, known as UB2, is not yet clear. However, DNA sequences recognized by UB2 were reported to colocalize with some of the known MARs.[84]

Calmodulin

Calmodulin, a protein present both in a cytoplasm and in nuclei, has been shown to interact in vivo with a heterologous MAR introduced into yeast cells.[85] In the same study it was demonstrated that another previously unknown 57 kDa nuclear polypeptide with an evolutionary conserved epitope found in mammalian cells also interacted in vivo with a human MAR introduced into yeast cells. The gene encoding the above protein (called SMI1) has been cloned and sequenced. In yeast cells, SMI1 is required both for the cell growth and the DNA replication.[85]

Histone H1 and HMG proteins

Histone H1 and HMG proteins were reported to bind MARs preferentially, as compared to total DNA.[35,86-88] The binding occurs in the DNA minor groove and can be competed by a minor groove-binding drug, distamycin.[35,86] HMG I and histone H1 have a common DNA binding domain known as "AT hook" which ensures preferential interactions of these proteins with the narrow minor groove of AT-rich DNA sequences.[89,90] The above domain has recently been successfully used for constructing an artificial protein able to bind MARs/SARs with the extremely high affinity.[59] Besides the ability to interact with a narrow minor groove on DNA, HMG1 can also recognize specifically the cruciform structures[91,92] that may be found frequently in MAR DNA. The interaction of HMG 1 with DNA cruciforms can stabilize these noncanonical DNA structures that are thought to play some role in replication initiation (see chapter 9).

The MAR-binding sites on the nuclear matrix are either heterogenous or are composed of different MAR-binding proteins because both characteristic properties of MARs (i.e., the narrow minor grove and the high base-unpairing potential) seem to be important for binding. Removing of the unwinding potential of short synthetic oligonucleotides also possessing narrow minor grooves significantly reduces their binding to the nuclear matrix but does not prevent it completely.[33,34]

Among other proteins presumed to interact with DNA at loop anchorage sites we can mention here the nuclear matrix proteins interacting with bent DNA[93-95] and proteins tightly bound to DNA. The latter constitute the most enigmatic group of nuclear matrix proteins. Over a long period of time it was reported repeatedly by different authors that there is a group of proteins very tightly bound to the eukaryotic DNA and that these proteins could not be released by treatment with SDS and standard deproteinization procedures.[96-105] Different molecular weights have been reported for the proteins tightly bound to DNA, the more frequent estimates being in the range of 50-70 kDa. As follows from the analysis of the distribution of these proteins between matrix-bound and cleaved-off DNA fragments, they are concentrated in the regions of DNA attachment to the nuclear matrix.[98,99,101,105] More recent data suggest that these proteins may participate in the attachment to the nuclear matrix of the transcriptionally active DNA fraction.[100,106,107] Some evidence has been presented in the literature indicating that the above proteins are bound to DNA covalently.[108] Furthermore, it was reported that covalent complexes of DNA with some nuclear matrix proteins might be obtained by incubation under a certain conditions of a mixture of cloned DNA fragments with isolated nuclear matrices. The proteins involved in the formation of the above complexes were found to be distinct from DNA topoisomerase II.[109] More than ten years ago the isolation of antibodies against several proteins tightly bound to DNA was described.[110] However, up to now none of the proteins tightly bound to DNA has been isolated and none of the genes encoding these proteins has been cloned. Hence, the whole story has to be further investigated.

6.4. COMPLEX ORGANIZATION OF DNA LOOP ANCHORAGE SITES: AN ATTEMPT OF SYNTHESIS

Summarizing the data discussed in the previous sections of this chapter we can say that:

1. No simple consensus sequence of the DNA loop anchorage site has been identified by any of the approaches used.
2. Two kinds of genomic elements have been assumed to participate in DNA loop anchorage at the nuclear matrix.

There is no common abbreviation for the first kind of the above elements. We shall call it nuclear matrix DNA which is a set of nuclease-protected DNA fragments that can be isolated from the nuclear matrix. Nuclear matrix DNA is composed of relatively

GC-rich simple DNA sequences with imperfect internal repeats, palindromes and more or less pronounced asymmetry in the distribution of purines and pyrimidines between the complementary DNA chains.

The second kind of genomic elements assumed to participate in DNA loop anchorage at the nuclear matrix are MARs/SARs. Typical MARs/SARs are relatively AT-rich and easily-melting DNA sequences characterized by the alternation of A and T residues in each strand,[34] as well as by the presence of ATATTT motifs[28] and short homopolymeric runs of A and T.[35] Several MARs/SARs-binding proteins have been described.[63,66-68,73-79] Some of these proteins are not, however, the exclusive components of the nuclear matrix.

The data obtained in experiments with topoisomerase II-mediated DNA loop excision suggest that the length of a DNA fragment interacting with the nuclear matrix (and hence preferentially accessible to the cleavage by high-salt-insoluble topoisomerase II) may be relatively long.[111,112] Although this fragment as a whole is not protected from nucleases (and as will be discussed in chapter 8 is even preferentially accessible to nucleases), some short pieces of it may be as well protected from nucleases by the interaction with the nuclear matrix proteins (not exclusively with topoisomerase II). These sites of interaction with the nuclear matrix proteins scattered along the loop anchorage region may correspond to the short nuclear matrix DNA fragments characterized by a degenerated sequence organization (section 6.1).

Our resent data[60] suggest that MARs/SARs may also constitute an essential component of the loop anchorage region. Importantly, it is not a sufficient component because a number of MARs/SARs can be found outside LARs in loop DNA. It seems that a complex and not yet properly characterized combination of MARs/SARs and other genomic elements including GC-rich simple DNA sequences, bent DNA[36,113-115] and recognition sites for different DNA-binding proteins[116] within the loop anchorage regions is necessary for maintaining the proper topological organization of chromosomal DNA. This complex organization of the DNA loop anchorage sites can be illustrated taking as an example a loop anchorage site located about 3 kb upstream to the first gene of the chicken α-globin genes cluster (Fig. 6.5). The region extending from 2.5-6 kb upstream to the α-globin genes cluster contains both a replication origin[117,118] and a permanent site of DNA attachment

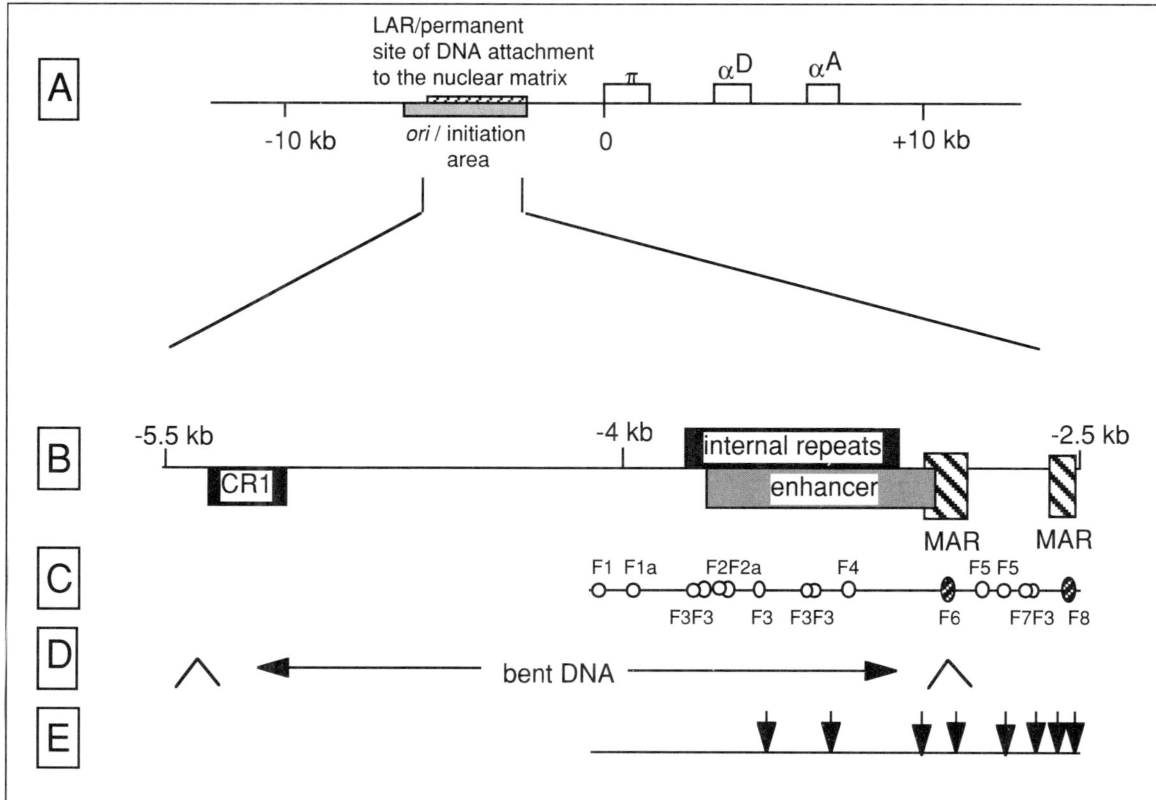

Fig. 6.5. A scheme illustrating complex organization of the DNA loop anchorage site in the upstream area of the chicken α-globin gene domain. (A) Low resolution map of the domain showing positions of genes (open rectangles) and position of the LAR/ori region. (B) A map of the LAR/ori region showing positions of CR1 repetitive element,[96] transcriptional enhancer,[105] area with internal repeats[104] and MAR elements.[103] (C) Positions of protein-binding sites within the right part of loop anchorage region. Designation of protein-binding sites (F1-F8) is given according to De Moura-Gallo et al.[106] F6 and F8 are recognition sites for the presumable replication initiation factors present only in extracts of proliferating cells.[106] F6 was found to be also a nuclear matrix protein.[108] Recognition site F4 is identical to one of the binding sequences of the c-myc protein. Distribution of protein-binding sites to the left of the recognition site for F1 was not studied. (D) Positions of bent DNA.[96] (E) Positions of preferential DNA cleavage in vitro by the topoisomerase II (determined only within the area shown by the horizontal line).[103]

to the nuclear matrix mapped by topoisomerase II-mediated DNA loop excision[119] and by isolation of nuclear matrix DNA from mature erythrocyte nuclei.[120,121] The scheme shown in Figure 6.5 represents only a part of the above region. However, even in this part one can find two MAR elements,[122] two permanently curved DNA sequences,[115] a G-C rich region with imperfect repeats[123] which possess an enhancer activity[124] and a number of recognition sites for sequence specific DNA binding proteins, including c-myc, GATA-1 and several novel motifs footprinted in vitro and in vivo.[125-127] The same region contains multiple recognition sites for in vitro DNA cleavage by DNA topoisomerase II.[122]

A considerable size and the complex organization of LARs can explain the failure to identify any matrix attachment consensus DNA sequence by analyzing short nuclear matrix DNA fragments isolated after limited nuclease digestion.

6.5. CONCLUSIONS

DNA loops are anchored to the nuclear matrix through relatively long DNA fragments which are likely to contain a specific set of sequence elements involved in the interaction with the nuclear matrix proteins. Among these elements there are MARs/SARs and also short DNA fragments with a degenerated sequence organization which can be isolated from the nuclear matrix extensively treated with nucleases. Any of the above sequence elements can be found also in loop DNA. Only in a certain combination they may form a loop anchorage region.

Related Review Articles

Blasquez VC, Sperry AO, Cockerill PN et al. Protein: DNA interactions at chromosomal loop attachment sites. Genome (Canada) 1989; 31:503-509.

Laemmli UK, Kas E, Poljak L et al. Scaffold-associated regions: cis-acting determinants of chromatin structural loops and functional domains. Curr Opin Genet Dev 1992; 2:275-285.

Boulikas T. Nature of DNA sequences at the attachment regions of genes to the nuclear matrix. J. Cell Biochem 1993; 52:14-22.

Boulikas T. Chromatin domains and prediction of MAR sequences. Int Rev Cytol 1995; 162A:279-388.

Razin SV, Gromova II, Iarovaia OV. Specificity and functional significance of DNA interaction with the nuclear matrix: New approaches to clarify the old questions. Int Rev of Cytol 1995; 162B:405-448.

REFERENCES

1. Razin SV, Mantieva VL, Georgiev GP. DNA adjacent to the attachment points of DNP fibril to chromosomal axial structure is enriched in reiterated base sequences. Nucl Acids Res 1978; 5:4737-4751.
2. Razin SV, Mantieva VL, Georgiev GP. The similarity of DNA sequences remaining bound to scaffold upon nuclease treatment of interphase nuclei and metaphase chromosomes. Nucl Acids Res 1979; 17:1713-1735.
3. Jeppesen PG, Bankier AT. A partial characterization of DNA fragments protected from nuclease degradation in histone depleted metaphase chromosomes of the Chinese hamster. Nucl Acids Res 1979; 7:49-67.
4. Kuo MT. Analysis of DNA attached to the chromosome scaffold. J Cell Biol 1982; 93:278-284.
5. Chimera JA, Musich PR. The association of the interspersed repetitive KpnI sequences with the nuclear matrix. J Biol Chem 1985; 260:9373-9379.
6. Matsumoto LH. Enrichment of satellite DNA on the nuclear matrix of bovine cells. Nature 1981; 294:481-482.
7. Basler J, Hastie ND, Pietras D et al. Hybridization of nuclear matrix attached deoxyribonucleic acid fragments. Biochemistry 1981; 20:6921-6929.
8. Bowen BC. DNA fragments associated with chromosomal scaffolds. Nucl Acids Res 1981; 19:5093-5108.
9. Jackson DA, Cook PR, Patel SB. Attachment of repeated sequences to the nuclear cage. Nucl Acids Res 1984; 17:6709-6726.
10. Small D, Nelkin B, Vogelstein B. Nonrandom distribution of repeated sequences with respect to supercoiled loops and the nuclear matrix. Proc Natl Acad Sci USA 1982; 70:5911-5915.
11. Kalandadze AG, Razin SV, Georgiev GP. Cloning and analysis of DNA fragments permanently attached to the nuclear matrix. Doklady AN SSSR (Moscow) 1988; 300:1001-1005.
12. Bogdanova AN, Razin SV, Vassetzky YS. Nuclear matrix-associated fragments enhance autonomous replication of plasmids in chicken cells. Biochimie 1995; 77:880-887.
13. Goldberg GI, Collier I, Cassel A. Specific DNA sequences associated with the nuclear matrix in synchronized mouse 3T3 cells. Proc Natl Acad Sci USA 1983; 80:6887-6891.
14. Opstelten RJG, Clement JME, Wanka F. Direct repeats at nuclear matrix-associated DNA regions and their putative control function in the replicating eukaryotic genome. Chromosoma 1989; 98:422-427.
15. Boulikas T, Kong CF. Multitude of inverted repeats characterizes a class of anchorage sites of chromatin loops to the nuclear matrix. J Cell Biochem 1993; 53:1-12.

16. Boulikas T, Kong CF. A novel class of matrix attached regions (MARs) identified by random cloning and their implication in differentiation and carcinogenesis. Int J Oncol 1993; 2:325-330.
17. Robinson SI, Nelkin BD, Vogelstein B. The ovalbumin gene is associated with the nuclear matrix of chicken oviduct cells. Cell 1982; 28:99-106.
18. Robinson SI, Small D, Idzerda R et al. The association of active genes with the nuclear matrix of the chicken oviduct. Nucl Acids Res 1983; 15: 5113-5130.
19. Ciejek EM, Tsai M-J, O'Malley BW. Actively transcribed genes are associated with the nuclear matrix. Nature 1983; 306:607-609.
20. Jost J-P, Seldran M. Association of transcriptionally active vitellogenin II gene with the nuclear matrix of chicken liver. EMBO J 1984; 3:2205-2208.
21. Small D, Vogelstein B. The anatomy of supercoiled loops in the *Drosophila* 7F locus. Nucl Acids Res 1985; 13:7703-7713.
22. Small D, Nelkin B, Vogelstein B. The association of transcribed genes with the nuclear matrix of *Drosophila* cells during heat shock. Nucl Acids Res 1985; 13:2413-2431.
23. Comings DE, Wallack AS. DNA-binding properties of nuclear matrix proteins. J Cell Sci 1978; 34:233-246.
24. Lesser BH, Comings DE. Biochim Biophys Acta 1978; 521: 117-125.
25. Nose K. Characterization of nuclear matrix from cultural normal human fibroblasts. J Biochem 1986; 99:1385-1391.
26. Mullenders LH, Eygenstein J, Broen A et al. Composition and DNA-binding properties of the nuclear matrix proteins. Biochim Biophis Acta 1982; 698:70-77.
27. Krachmarov C, Iovcheva C, Hancock R et al. Association of DNA with the nuclear lamina in Ehrlich ascites tumor cells. J Cell Biochem 1986; 31:59-74.
28. Cockerill PN, Garrard WT. Chromosomal loop anchorage of the kappa immunoglobulin gene occurs next to the enhancer in a region containing topoisomerase II sites. Cell 1986; 44:273-282.
29. Cockerill PN, Garrard WT. Chromosomal loop anchorage sites appear to be evolutionary conserved. FEBS Lett 1986; 204:5-7.
30. Sperry AO, Blasquez VB, Garrard WT. Dysfunction of chromosomal loop attachment sites: illegitimate recombination linked to matrix associated regions and topoisomerase II. Proc Natl Acad Sci USA 1989; 86:5497-5501.
31. Amati B, Gasser SM. *Drosophila* scaffold-attached regions bind nuclear scaffolds and can function as ARS elements in both budding and fission yeasts. Mol Cell Biol 1990; 10:5442-5454.
32. Das AT, Luderus ME, Lamers WH. Identification and analysis of a matrix-attachment region 5' of the rat glutamate-dehydrogenase-encoding gene. Eur J Biochem 1993; 215:777-785.

33. Mielke C, Kohwi I, Kohwi-Shigematsu T et al. Hierarchical binding of DNA fragments derived from scaffold-attached regions: correlation of properties in vitro and function in vivo. Biochemistry 1990; 29:7475-7485.
34. Bode J, Kohwi I, Dickinson L et al. Biological significance of unwinding capability of nuclear matrix-associated DNAs. Science 1992; 255:195-197.
35. Kas E, Izaurralde E, Laemmli UK. Specific inhibition of DNA binding to nuclear scaffolds and histone H1 by distamycin: The role of oligo(dA)·oligo(dT) tracts. J Mol Biol 1989; 210:587-599.
36. Homberger HP. Bent DNA is a structural feature of scaffold-attached regions in *Drosophila melanogaster* interphase nuclei. Chromosoma 1989; 98:99-104.
37. Mielke C, Maass K, Tummler M, Bode J. Anatomy of highly expressed chromosomal sites targeted by retroviral vectors. Biochemistry 1996; 35:2239-2252.
38. Cockerill PN, Yuen M-H, Garrard WT. The enhancer of the immunoglobulin heavy chain locus is flanked by presumptive chromosomal loop anchorage elements. J Biol Chem 1987; 262: 5394-5397.
39. Dijkwel PA, Hamlin JL. Matrix attachment regions are positioned near replication initiation sites, genes, and an interamplicon junction in the amplified dihydrofolate reductase domain of Chinese hamster ovary cells. Mol Cell Biol 1988; 8:5398-53409.
40. Phi-Van L, Stratling WH. The matrix attachment regions of the chicken lysozyme gene co-map with the boundaries of the chromatin domain. EMBO J 1988; 7:655-664.
41. Bode J, Maass K. Chromatin domain surrounding the human interferon-β gene as defined by scaffold attached regions. Biochemistry 1988; 27:4706-4711.
42. Levy-Wilson B, Fortier C. The limits of the DNase I-sensitive domain in the human apolipoprotein B gene coincide with the locations of chromosomal anchorage loops and define the 5' and 3' boundaries of the gene. J Biol Chem 1989; 264:21196-21204.
43. Ito T, Sakaki Y. Nuclear matrix association regions of rat alpha 2-macroglobulin gene. Biochem Biophys Res Commun 1987; 149:449-454.
44. Kas E, Chasin LA. Anchorage of the Chinese hamster dihydrofolate reductase gene to the nuclear scaffold occurs in an intragenic region. J Mol Biol 1987; 198:577-692.
45. Beggs AH, Migeon BR. Chromatin loop structure of the human X chromosome: Relevance to X inactivation and CpG clusters. Mol Cell Biol 1989; 9:2322-2331.
46. Romig H, Ruff J, Fackelmayer FO et al. Characterization of two intronic nuclear-matrix-attachment regions in the human DNA topoisomerase I gene. Eur J Biochem 1994; 221:411-419.

47. Hakes DJ, Berezney R. DNA binding properties of the nuclear matrix and individual nuclear matrix proteins. Evidence for salt-resistant DNA binding sites. J Biol Chem 1991; 266:11131-11140.
48. Kay V, Bode J. Binding specificity of a nuclear scaffold: Supercoiled, single-stranded and scaffold-attached-region DNA. Biochemistry 1994; 33:367-374.
49. Tsutsui K, Tsutsui K, Muller MT. The nuclear scaffold exhibits DNA-binding sites selective for supercoiled DNA. J Biol Chem 1988; 263:7235-7241.
50. Gasser SM, Amati BB, Cardenas ME et al. Studies on scaffold attachment sites and their relation to genome function. Int Rev Cytol 1989; 119:57-96.
51. Mirkovich J, Mirault E, Laemmli UK. Organization of the higher order chromatin loop: specific DNA attachment sites on nuclear scaffold. Cell 1984; 39:323-332.
52. Cook PR, Brazell IA. Mapping sequences in loops of nuclear DNA by their progressive detachment from the nuclear cage. Nucl Acids Res 1980; 8:2895-2906.
53. Robinson SI, Nelkin BD, Vogelstein B. The ovalbumin gene is associated with the nuclear matrix of chicken oviduct cells. Cell 1982; 28:99-106.
54. Robinson SI, Small D, Idzerda R et al. The association of active genes with the nuclear matrix of the chicken oviduct. Nucl Acids Res 1983; 15:5113-5130.
55. Izaurralde E, Mirkovich J, Laemmli UK. Interaction of DNA with nuclear scaffolds in vitro. J Mol Biol 1988; 200:111-125.
56. Mirkovich J, Gasser SM, Laemmli UK. Scaffold attachment of DNA loops in Metaphase chromosomes. J Mol Biol 1988; 200:101-109.
57. Eggert H, Jack RS. An ectopic copy of the *Drosophila* ftz associated SAR neither recognizes local chromatin structure nor hinders elution of a chromatin fragment from isolated nuclei. EMBO J 1991; 10:1237-1243.
58. Hempel K, Stratling WH. The chicken lysozyme gene 5' MAR and the *Drosophila* histone SAR are electroelutable from encapsulated and digested nuclei. J Cell Sci 1996; 109:(in press).
59. Strick R, Laemmli UK. SARs are cis DNA elements of chromosome dynamics: Synthesis of a SAR repressor protein. Cell 1996; 83:1137-1148.
60. Iarovaia OV, Hancock R, Lagarkova MA et al. Mapping of genomic DNA loop organization in a 500-kilobase region of the *Drosophila* X chromosome using the topoisomerase II-mediated DNA loop excision protocol. Mol Cell Biol 1996; 16:302-308.
61. Surdej P, Got C, Rosset R et al. Supragenic loop organization: mapping in *Drosophila* embryos, of scaffold-associated regions on a 800 kilobase continuum cloned from the 14B-15B first chromosome region. Nucl Acids Res 1990; 18:3713-3722.

62. Brun C, Dang Q, Miassod R. Studies of an 800-kilobase DNA stretch of the *Drosophila* X chromosome: comparing of a subclass of scaffold-attached regions with sequences able to replicate autonomously in Saccharomyces cerevisiae. Mol Cell Biol 1990; 10: 5455-5463.
63. Von Kries JP, Buhrmester H, Stratling WH. A matrix/scaffold attachment region binding protein: identification, purification, and mode of binding. Cell 1991; 64:123-135.
64. Kaufmann SH, Coffey DS, Shaper JH. Considerations in the isolation of rat liver nuclear matrix, nuclear envelope, and pore complex lamina. Exp Cell Res 1981; 132:105-123.
65. Kaufmann SH, Gibson W, Shapper JH. Characteization of the major polypeptides of the rat liver nuclear envelope. J Biol Chem 1983; 258:2710-2719.
66. Buhrmester H, Von Kries JP, Stratling WH. Nuclear matrix protein ARBP recognizes a novel DNA sequence motif with high affinity. Biochemistry 1994; 34:4108-4117.
67. Romig H, Fackelmayer FO, Renz A et al. Characterization of SAF-A, a novel nuclear DNA binding protein from HeLa cells with high affinity for nuclear matrix/scaffold attachment elements. EMBO J 1992; 11:3431-3440.
68. Tsutsui K, Tsutsui K, Okada S et al. Identification and characterization of a nuclear scaffold protein that binds the nuclear matrix attachment region DNA. J Biol Chem 1993; 268:12886-12894.
69. Hofmann JF-X, Laroche T, Brand AH et al. RAP-1 factor is necessary for DNA loop formation in vitro at the silent mating type locus HML. Cell 1988; 57:725-737.
70. Fackelmayer FO, Dahm K, Renz A et al. Nucleic-acid-binding properties of hnRNP-U/SAF-A, a nuclear matrix protein which binds DNA and RNA in vivo and in vitro. Eur J Biochem 1994; 221:749-757.
71. Von Kries JP, Buck F, Stratling WH. Chicken MAR binding protein p120 is identical to human heterogeneous nuclear ribonucleoprotein (hnRNP) U. Nucl Acids Res 1994; 22:1215-1220.
72. Renz A, Fackelmayer FO. Purification and molecular cloning of the scaffold attachment factor B (SAF-B), a novel human nuclear protein that specifically binds to S/MAR-DNA. Nucl Acids Res 1996; 24:843-849.
73. Luderus ME, de Graaf A, Mattia E et al. Binding of matrix attachment regions to lamin B1. Cell 1992; 70:949-959.
74. Luderus ME, den Blaauwen JL, de Smit OJ et al. Binding of matrix attachment regions to lamin polymers involves single-stranded regions and the minor groove. Mol Cell Biol 1994; 14:6297-6305.
75. Zhao K, Harel A, Stuurman N et al. Binding of matrix attachment regions to nuclear lamin is mediated by the rod domain and depends on the lamin polymerization state. FEBS Lett 1996; 380:161-164.

76. Dickinson LA, Joh T, Kohwi Y et al. A tissue-specific MAR/SAR DNA-binding protein with unusual binding site recognition. Cell 1992; 70:631-645.
77. Nakagomi K, Kohwi Y, Dickinson LA et al. A novel DNA-binding motif in the Nuclear Matrix Attachment DNA binding protein SATB1. Mol Cell Biol 1994; 14:1852-1860.
78. Kohwi-Shigematsu T, Kohwi Y. Torsional stress stabilizes extended base unpairing in suppressor sites flanking immunoglobulin heavy chain enhancer. Biochemistry 1990; 29:9551-9560.
79. Herrscher RF, Kaplan MH, Lelsz DL et al. The immunoglobulin heavy-chain matrix-associating regions are bound by Bright: a B cell-specific trans-activator that describes a new DNA-binding protein family. Genes Dev 1996; 9:3067-3082.
80. Yanagisawa J, Ando J, Nakayama J et al. A matrix attachment region (MAR)-binding activity due to a p114 kilodalton protein is found only in human breast carcinomas and not in normal and benign breast disease tissues. Cancer Res 1996; 56:457-462.
81. Wang B, Dickinson LA, Koivunen E et al. A novel matrix attachment region DNA binding motif identified using a random phage peptide library. J Biol Chem 1995; 270;23239-23242.
82. Dickinson LA, Kohwi-Shigematsu T. Nucleolin is a matrix attachment region DNA-binding protein that specifically recognizes a region with high base-unpairing potential. Mol Cell Biol 1995; 15:456-465.
83. Adachi Y, Kas E, Laemmli UK. Preferential, cooperative binding of DNA topoisomerase II to scaffold-associated regions. EMBO J 1989; 8:3997-4006.
84. Ma X, Saitoh N, Curtis PJ. Purification and characterization of a nuclear DNA-binding factor complex containing topoisomerase II and chromosome scaffold protein 2. J Biol Chem 1993; 268: 6182-6188.
85. Fishel BR, Sperry AO, Garrard WT. Yeast calmodulin and a conserved nuclear protein participate in the in vivo binding of matrix associated region. Proc Natl Acad Sci USA 1993; 90:5623-5627.
86. Izauralde E, Kas E, Laemmli UK. Highly preferential nucleation of histone H1 assembly on scaffold-associated regions. J Mol Biol 1989; 210:573-585.
87. Ivanchenko M, Avramova Z. Interaction of MAR-sequences with nuclear matrix proteins. J Cell Biochem 1992; 50:190-200.
88. Zhao K, Kas E, Gonzalez E et al. SAR-dependent mobilization of histone H1 by HMG-I/Y in vitro: HMG-I/Y is enriched in H1-depleted chromatin. EMBO J 1993; 12:3237-3247.
89. Reeves R, Nissen MS. The A·T-DNA-binding domain of mammalian high mobility group I chromosomal proteins: A novel peptide motif for recognizing DNA structure. J Biol Chem 1990; 265:8573-8582.

90. Reeves R, Langan TA, Nissen MS. Phosphorilation of the DNA-binding domain of nonhistone high-mobility group I protein by cdc2 kinase: Reduction of binding affinity. Proc Natl Acad Sci USA 1991; 88:1671-1675.
91. Bianchi ME, Beltrame M, Paonessa G. Specific recognition of cruciform DNA by nuclear protein MHG1. Science 1989; 243: 1056-1059.
92. Bianchi ME, Falciola L, Ferrari S et al. The DNA binding site of HMG1 protein is composed of two similar segments (HMG boxes), both of which have counterparts in other eukaryotic regulatory proteins. EMBO J 1992; 11:1055-1063.
93. Hibino Y, Nakamura K, Asano S et al. Affinity of highly repetitive bent DNA for nuclear scaffold proteins from rat liver. Biochem Biophys Res Commun 1992; 184:853-858.
94. Hibino Y, Nakamura K, Tsukada S et al. Purification and characterization of nuclear scaffold proteins which bind to a highly repetitive bent DNA from rat liver. Biochim Biophys Acta 1993; 1174:162-170.
95. Hibino Y, Fujii K, Tsukada, S et al. Binding affinities of highly repetitive DNA components for a nuclear scaffold protein from rat-ascite hepatoma cells. Cancer Lett 1995; 88:49-55.
96. Krauth W, Werner D. Analysis of the most tightly bound proteins in eukaryotic DNA. Biochim Biophys Acta 1979; 564:390-401.
97. Werner D, Petzelt C. Alkali-stably bound proteins in eukaryotic and prokaryotic DNAs show common characteristics. J Mol Biol 1981; 150:297-302.
98. Razin SV, Chernokhvostov VV, Roodin AV et al. Proteins tightly bound to DNA in the regions of DNA attachment to the skeletal structures of interphase nuclei and metaphase chromosomes. Cell 1981; 27:65-73.
99. Bodnar JW, Jones CJ, Coombs DH et al. Proteins tightly bound to HeLa cell DNA at nuclear matrix attachment sites. Mol Cell Biol 1983; 3:1567-1579.
100. Neuer B, Werner D. Screening of isolated DNA for sequences released from anchorage sites in the nuclear matrix. J Mol Biol 1985; 181:15-25.
101. Neuer-Nitsche B, Werner D. Sub-set characteristics of DNA sequences involved in tight DNA/polypeptide complexes and their homology to nuclear matrix DNA. Biochem. Biophys Res Commun 1987; 147:335-339.
102. Neuer-Nitsche B, Lu XN, Werner D. Functional role of a highly repetitive DNA sequence in anchorage of the mouse genome. Nucl Acids Res 1988; 16:8351-8360.
103. Avramova Z, Tsanev R. Stable DNA-protein complexes in eukaryotic chromatin. J Mol Biol 1987; 196:437-440.

104. Avramova Z, Mikhailov I, Tsanev R. An evolutionary conserved protein fraction stably linked to DNA. Biochim Biophys Acta 1989; 1007:109-111.
105. Werner D, Rest R. Radiolabeling of DNA/polypeptide complexes in isolated bulk DNA and in residual nuclear matrix DNA by nick-translation. Biochem Biophys Res Commun 1987; 147:340-345.
106. Razin SV, Chernokhvostov VV, Vassetzky YS et al. The distribution of tightly bound proteins along the DNA chain reflects the pattern of cell differentiation. Nucl Acids Res 1988; 16:3617-3634.
107. Leibovitch SA, leibovitch MP, Hillion H et al. A destabilized DNA conformation associated with tightly bound nuclear proteins in active genes of rat myoblast. Nucl Asids Res 1983; 11:4035-4047.
108. Neuer B, Plagens U, Werner D. Phosphodiester bounds between polypeptides and chromosomal DNA. J Mol Biol 1983; 164: 213-235.
109. Zenk DW, Ginder GD, Brotherton TW. A nuclear matrix protein binds tightly to DNA in the avian beta-globin gene enhancer. Biochemistry 1990; 29:5221-5226.
110. Werner D, Chanpu S, Muller M et al. Antibodies to the most tightly bound proteins in eukaryotic DNA. Formation of immunocomplexes with 'nuclear matrix' components. Exp Cell Res 1984; 151:384-395.
111. Razin SV, Hancock R, Iarovaia O et al. Structural-functional organization of chromosomal DNA domains. Cold Spring Harb Symp Quant Biol 1993; 58:25-35.
112. Gromova II, Thomsen B, Razin SV. Different topoisomerase II antitumour drugs direct similar specific long-range fragmentation of an amplified *c-myc* gene locus in living cells and in high-salt-extracted nuclei. Proc Natl Acad Sci USA 1995; 92:102-106.
113. Anderson JN. Detection, sequence patterns and function of unusual DNA structures. Nucl Acids Res 1986; 14:8513-8533.
114. Von Kries JP, Phi-Van L, Diekmann S et al. A non-curved chicken lysozyme 5' matrix attachment site is 3' followed by a strongly curved DNA sequence. Nucl Acids Res 1990; 18:3881-3885.
115. Krajewski WA, Razin SV. Organization of specific DNA sequence elements in the region of replication origin and matrix attachment site of the chicken alpha-globin gene domain. Mol Gen Genet 1992; 235:381-388.
116. Boulikas T. Transcription factor binding sites in the matrix attachment region (MAR) of the chicken alpha-globin gene. J Cell Biochem 1994; 55:513-529.
117. Razin SV, Kekelidze MG, Lukanidin EM et al. Replication origins are attached to the nuclear skeleton. Nucl Acids Res 1986; 14:8189-8207.
118. Verbovaia L, Razin SV. Analysis of the replication direction through the domain of α-globin-encoding genes. Gene 1995; 166:255-259.

119. Razin SV, Petrov P, Hancock R. Precise localization of the α-globin gene cluster within one of 20 to 300 kb fragments released by cleavage of chicken chromosomal DNA at topoisomerase II sites in vivo: evidence that the fragments are DNA loops or domains. Proc Natl Acad Sci USA 1991; 88:8515-8519.
120. Razin SV, Rzeszowska-Wolny J, Moreau J et al. Localization of regions of DNA attachment to the nuclear skeleton within chicken alpha-globin genes in functionally active and functionally inactive nuclei. Mol Biol (Moscow) 1985; 19:456-466.
121. Farache G, Razin SV, Rzeszowska-Wolny J et al. Mapping of structural and transcription-related matrix attachment sites in the α-globin gene domain of avian erythroblasts and erythrocytes. Mol Cell Biol 1990; 10:5349-5358.
122. Razin SV, Vassetzky YS, Hancock R. Nuclear matrix attachment regions and topoisomerase II binding and reaction sites in the vicinity of a chicken DNA replication origin. Biochem Biophys Res Commun 1991; 177:265-270.
123. Kalandadze AG, Bushara SA, Vassetzky YS et al. Characterization of DNA pattern in the site of permanent attachment to the nuclear matrix located in the vicinity of replication origin. Biochem Biophys Res Commun 1990; 168:9-15.
124. Razin SV, Vassetzky Jr YS, Kvartskhava AI et al. Transcriptional enhancer in the vicinity of a replication origin within the 5'-region of the chicken α-globin gene domain. J Mol Biol 1991; 217: 595-598.
125. De Moura-Gallo CV, Vassetzky YS, Recillas Targa F et al. The presence of sequence-specific protein binding sites correlate with replication activity and matrix binding in a 1.7 Kb-long DNA fragment of the chicken a-globin gene domain. Biochem Biophys Res Commun 1991; 179:512-519.
126. Recillas-Targa F, Razin SV, De Moura Gallo CV et al. Excision close to matrix attachment regions of the entire chicken α-globin genes domain by nuclease S1 and characterization of framing structures. Proc Natl Acad Sci USA 1994; 91:4422-4426.
127. Vassetzky YS, De Moura Gallo CV, Bogdanova AN et al. The sequence-specific nuclear matrix-binding factor F6 is a chicken GATA-like protein. Mol Gen Genet 1993; 238:309-314.

CHAPTER 7

Transient Interactions of DNA with the Nuclear Matrix

7.1. ASSOCIATIONS OF DNA WITH THE NUCLEAR MATRIX ORIGINATING IN CONNECTION WITH DNA REPAIR AND REPLICATION

DNA directly participates in three major functional processes: replication, repair of DNA damages and transcription. Although all these processes were reported to occur at the nuclear matrix,[1-3] little is known about the organization of function-dependent anchorage of DNA at the nuclear matrix. As far as DNA repair is concerned, the whole matter is still extremely controversial. In some cases, no preferential DNA repair at the nuclear matrix was observed.[4-6] Nevertheless, two important repair enzymes (Poly (ADP-ribose) polymerase and DNA polymerase β) have been localized on the nuclear matrix by immunochemical approaches.[7,8] Whatever a true organization of different types of DNA attachments to the nuclear matrix may be, it would be reasonable to assume that all these associations are maintained via the interaction of DNA with the nuclear matrix proteins. Hence, the repairing DNA sequences may be transiently bound to the nuclear matrix through the above mentioned enzymes.

The results of experiments demonstrating replication-dependent association of DNA with the nuclear matrix have been discussed in chapter 4. Many enzymes involved in DNA replication were found to reside in the nuclear matrix compartment. These are DNA polymerase α,[9-11] DNA primase,[12-14] proliferating cell nuclear antigen (PCNA)[15] and RNase H.[16] It is most likely that all these enzymes are assembled into huge replication complexes.[17,18] The latter are bound to the nuclear matrix,[17] and replication proceeds while DNA moves through these complexes. The topological models describing the possibility of chromosomal DNA replication by immobilized replication complexes were discussed previously by P. Cook.[19] Recent evidence indicates that individual replication complexes are combined into large megacomplexes. After immunostaining of nascent DNA, these megacomplexes can be visualized as replication foci at the level of light microscope resolution.[20-22] The integrity of replication foci is not affected by high-salt extraction

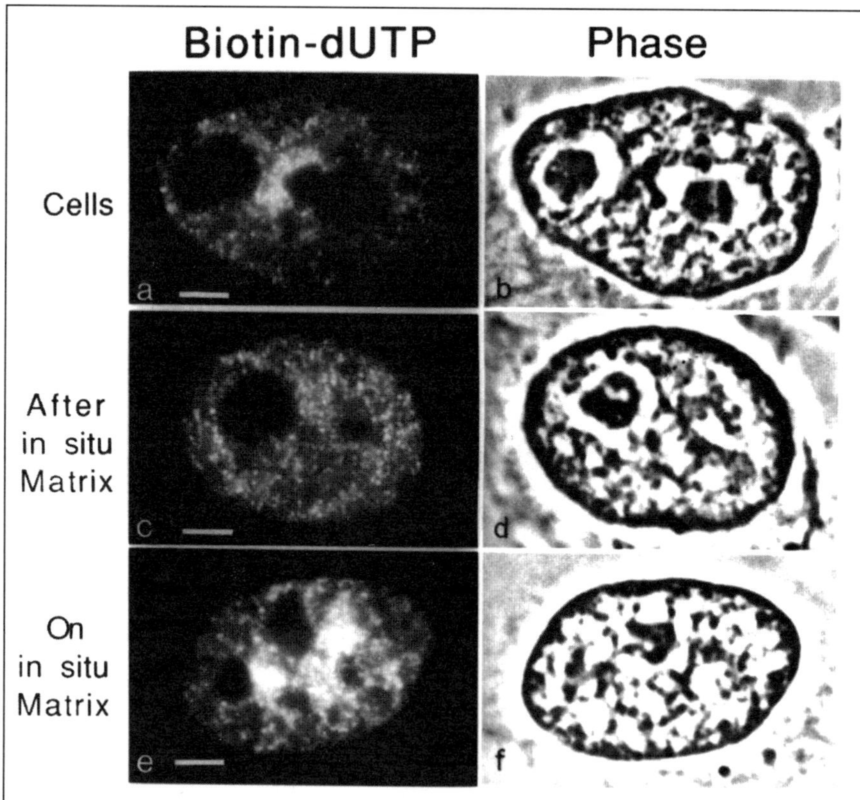

Fig. 7.1. DNA synthesis in permeabilized cells and after extraction for in situ nuclear matrix. (a and b) PtK-1 cells on coverslips were permeabilized with 0.04% Triton X-100 and incubated with the DNA synthesis medium at 37°C for 5 min. In c and d, the cells were then treated with DNase I and 0.2 M ammonium sulfate to prepare in situ nuclear matrix structures. In e and f, in situ nuclear matrices were first prepared followed by the incubation for DNA synthesis. Bars, 4 µm. (Reprinted with permission of The Rockefeller University Press from Nakayasu H, Berezney R. J Cell Biol 1989; 108:1-11. © 1989 by The Rockefeller University Press.)

and they retain the ability to incorporate precursors into DNA after isolation of nuclear matrices[20] (Fig. 7.1). The replication foci most likely correspond to the replication factories observed by Cook and collaborators on electron microscopic pictures of thick resinless sections of chromatin-depleted nuclei.[23,24] The electron-dense replication factories are characterized by a distinct (ovoid-like) shape and a relatively constant size (100-300 nm). They are scattered over the whole nucleus being attached to the diffuse nuclear skeleton. Using pulse-labeling with biotin-11-dUTP, Cook and collaborators have demonstrated that the ovoid bodies are the places of active DNA synthesis[23] (Fig. 7.2).

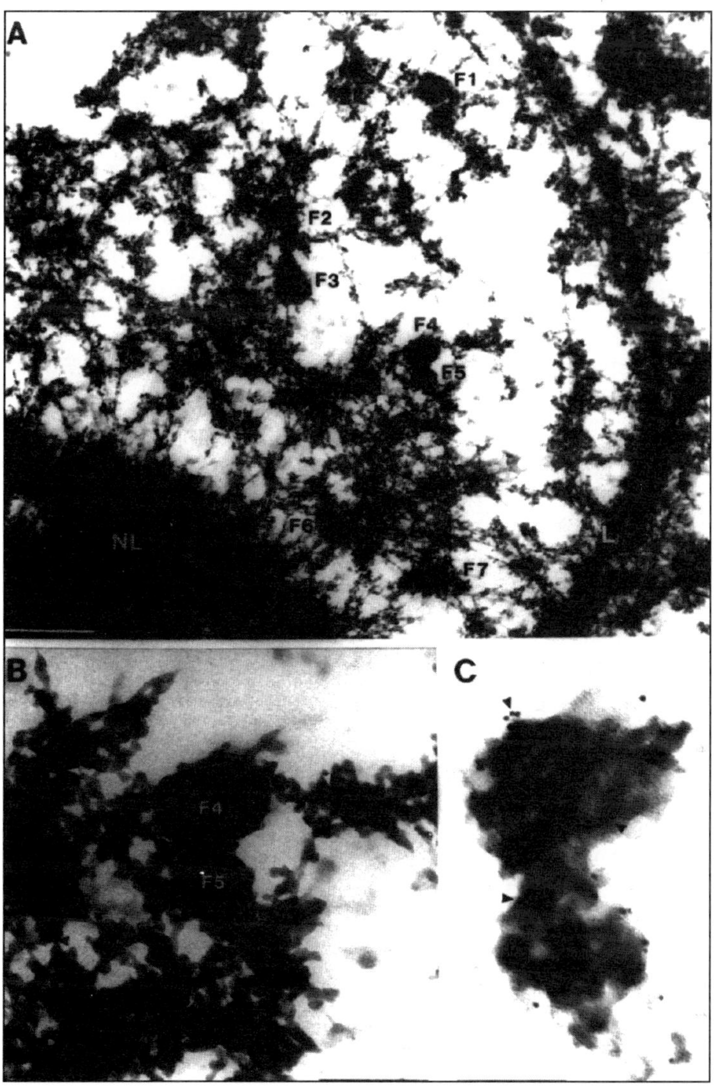

Fig. 7.2. Replication factories pulse-labeled with biotin-11-dUTP. Encapsulated cells were permeabilized with streptolysin O, incubated with biotin-11-dUTP for 2.5 min and treated with nucleases; about 90% of the chromatin was eluted; and sites of biotin incorporation were immunolabeled. (A) Medium power view; NL, nucleolus; L, nuclear lamina; F1-F7, ovoid bodies (i.e., replication factories). Gold particles cannot be seen at this magnification. (B) Higher power view of F4 and F5 in (A). (C) Underexposure and 2x magnification of (B) to show labeling. Three arrowheads indicate some of 30 gold particles in F4 and F5. (Reprinted with permission of Cell Press from Hozak P, Hassan AB, Jackson DA, Cook PR. Cell 1993; 73:361-373. © 1993 by Cell Press.)

Multienzyme complexes involved in DNA replication can be released from the nuclear matrix as 100S-150S conglomerates (clusterosomes) by the mild sonication procedure. In the course of in vitro incubation of matrix extracts these megacomplexes tend to dissociate into 10S and 17S subunits[17] (Fig. 7.3). Simple calculations show that 100S-150S clusterosomes may be composed of 10 to 30 small (10S) complexes.[17] This figure is similar to the estimated number of replicons in replicon clusters which may be visualized by fiber autoradiography.[25-27] It is of interest that the possibility to isolate from nuclear matrices the 100S-150S conglomerates of replication complexes strictly depends on cell proliferation status.[17] It is likely that these conglomerates are assembled only during active DNA replication. In contrast, the 10S replication complexes are assembled before the onset of replication (refs. 9, 10, 17 and Fig. 7.3).

7.2. INTERACTIONS OF TRANSCRIPTIONALLY ACTIVE DNA WITH THE NUCLEAR MATRIX

In chapter 4, the data have been discussed which indicate that transcription per se is not a prerequisite for apparent association of active genes with the nuclear matrix.[28-30] However, in all the above cited works the resolution of mapping experiments was relatively low. Consequently, it was not possible to discriminate an association with the nuclear matrix of the upstream regulatory sequences (including a promoter) and of the body of the gene (for a discussion see refs. 31, 32). Presently it seems likely that upstream regulatory regions of potentially transcribing genes are bound to the nuclear matrix due to the specific interaction of matrix-bound transcription factors[33-38] with target sequences within the above regions (Fig. 7.4). These interactions create a proper three-dimensional organization of the upstream regulatory regions. The rapid progress in the identification of different factors regulating expression of almost each gene clearly demonstrates the necessity of a three-dimensional organization which can bring and hold together distant regulatory elements (such as Locus Control Region, enhancer and promoter) by looping out the spacer DNA fragments (Fig. 7.4) and hence facilitate mutual interactions between these elements necessary for their function[39] (for a discussion see refs. 40-42). It should probably be underlined here that at least some

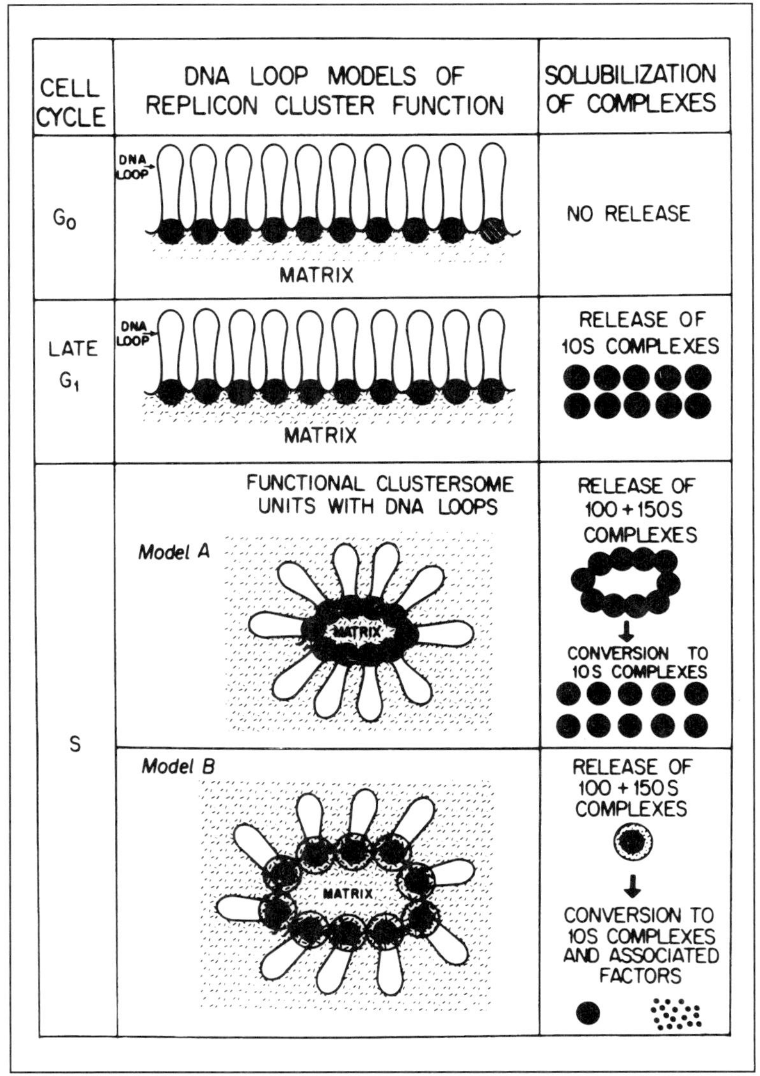

Fig. 7.3. Properties of nuclear matrix-released α-polymerase-primase complexes and their possible role in DNA replication. G_0 represents normal rat liver; late G_1 represents prereplicative regenerating rat liver (14 h posthepatectomy); and S represents regenerating rat liver (22 h posthepatectomy). The nuclear matrix of normal liver is virtually devoid of DNA polymerase α and primase activity. Therefore none of these activities is detected in the matrix extract of normal rat liver. The matrix attachment sites devoid of replicative activities are shown by hatched black circles. By late G_1 there is significant assembly of α-polymerase and primase on the nuclear matrix which are released as predominantly 10 S complexes and depicted on the matrix attachment sites by solid black circles. During active replication, predominantly 100 and 150 S megacomplexes are released and DNA loops are presumed to be organized into functional replicon clusters. Two possible models of megacomplexes are presented. In model A, assembly of a replicon cluster is mediated by corresponding clustering of individual 10 S and/or 17 S complexes into a clustersome. Supporting this model is the complete in vitro dissociation of the 100 and 150 S megacomplexes into 10 and 17 S complexes. In model B, the prereplicative 10 and 17 S complexes represent the core structures (black centers of stippled white circles) of much larger individual replisomes (the 100 and 150 S megacomplexes) which are assembled during active DNA replication. The putative clustersome (replication focus) then corresponds to many of these megareplisomes. (Reprinted with permission of The American Society for Biochemistry and Molecular Biology from Tubo RA, Berezney R. J Biol Chem 1987; 262:5857-5865. © 1987 by the American Society of Biological Chemists, Inc.)

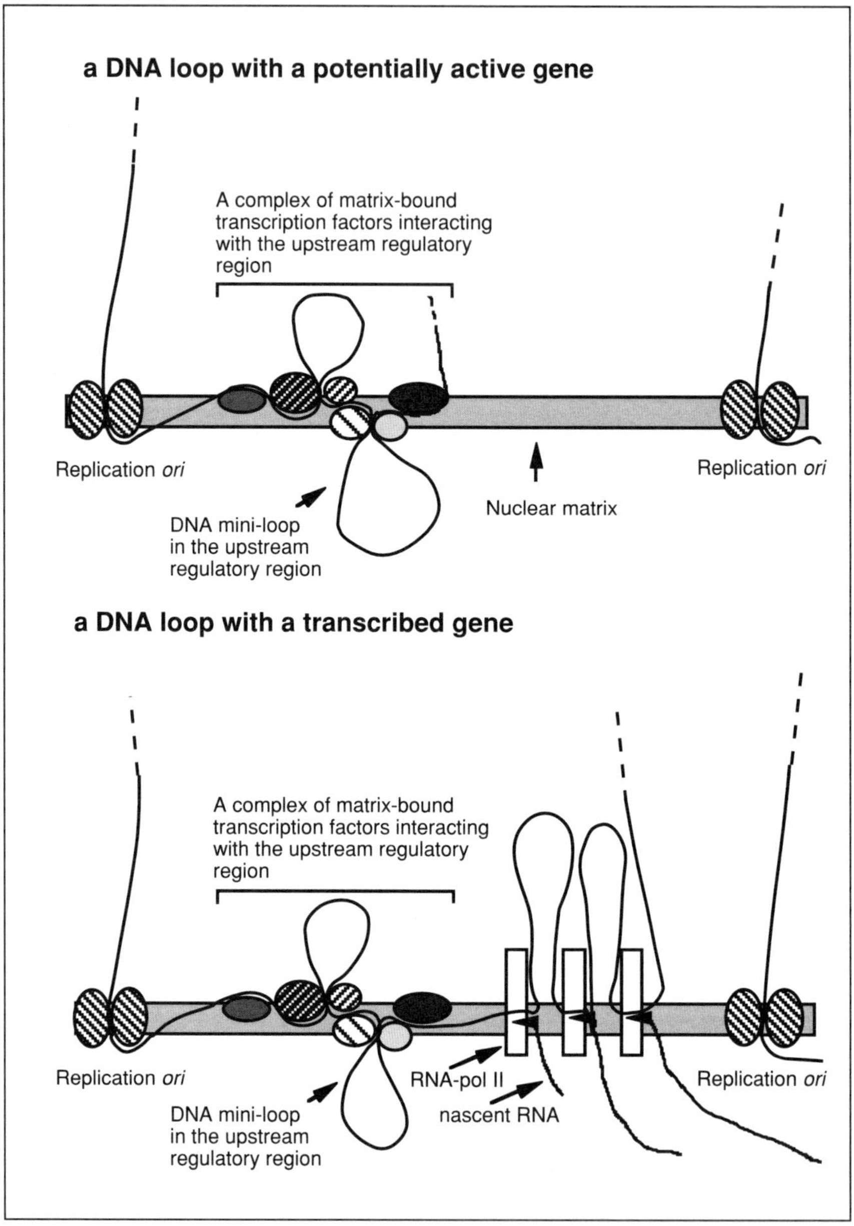

Fig. 7.4. A scheme illustrating different kinds of associations of potentially active and active genes with the nuclear matrix. (A) Within a basic loop defined by the attachment to the nuclear matrix of replication origins an upstream regulatory region of a potentially active gene is additionally attached to the nuclear matrix by sequence-specific interactions of target DNA sequences with matrix-bound transcription factors. Two relatively long regions of DNA which are located between protein-binding sites are looped out. (B) After start of transcription the "body" of a gene becomes transiently attached to the nuclear matrix via matrix-bound transcription complexes.

(and maybe most) of hormone receptors directly interact with target DNA sequences located within upstream regulatory domains of different tissue-specific genes (for examples see refs. 43-48). The receptors themselves and receptor-binding factors have been found to reside in the internal nuclear matrix. The spectrum of hormone receptors and receptor-binding factors represented in the nuclear matrix is clearly tissue-specific (for examples and discussion see chapter 1 and also ref. 49). Hence, the above interactions may be responsible for the attachment to the nuclear matrix of tissue-specific genes in cells undergoing differentiation in response to hormone treatment.

Interactions of upstream regulatory regions of potentially transcribing genes with the nuclear matrix do not depend on the transcription per se and may hold potentially active genes on the surface of transcription/splicing speckles. Indeed, according to recent observations the localization of several active genes on the surface of the transcription/splicing speckles is not affected by inhibition of transcription.[50] At the same time, nonactive genes do not colocalize with the above speckles.[50] Thus, the localization of genes in respect to transcription/splicing speckles (which constitute a part of the internal nuclear matrix) may be subjected to differentiation-dependent changes, although it is not affected by a simple arrest of transcription. One can see that these recent observations are in perfect agreement with those made in the initial experiments with high-salt-extracted nuclei.[28-30]

A relatively steady association of upstream regulatory sequences of transcriptionally active genes with the nuclear matrix does not exclude the possibility that transcribed regions may become additionally bound to the nuclear matrix through RNA polymerase in the course of transcription (Fig. 7.4B). This possibility is corroborated by two lines of experimental evidence. First, it has been found that truly transcribed genes are attached to the nuclear matrix by means of multiple interactions distributed within the whole transcription unit.[51] Second, RNA polymerize II elongation complexes were found in association preferentially (if not exclusively) with short fragments of nuclear matrix DNA.[52,53] Similar observations were made when loop DNA was electroeluted from nuclei in a "physiological ionic strength buffer." Active RNA polymerases (both RNA polymerase I and RNA polymerase II) resided in the nuclear matrix.[54,55] These results could easily be explained by a model postulating that an RNA polymerase (along with auxiliary proteins) is fixed at the nuclear matrix and that transcribed DNA moves through the transcription complexes in the course of transcription. This interpretation is further supported by the results of a model experiment with transcriptionally active minichromosomes.[56] Transcribed minichromosomes became transiently attached to the nuclear matrix. Following treatment with Hae III restriction endonuclease the major part of minichromosomal DNA could be eluted from nuclei. After elution of cleaved-off DNA fragments, the matrix-bound DNA was not represented by any particular region of the plasmid. All parts of transcriptional units could be found in this DNA fraction. These observations suggested that the association of minichromosomes with the nuclear matrix was mediated by RNA-polymerase II.[56] At first sight it seems difficult to understand how transcription can occur while the DNA template is passing through the RNA polymerase II immobilized at the nuclear matrix. However, elegant experiments of Cook and collaborators have demonstrated that an RNA polymerase immobilized (through a peptide linker) on plastic beads is able to accomplish transcription with about the same speed as the free enzyme in solution.[57] As a matter of fact, the model postulating that DNA moves through immobilized transcriptional complexes allows one logical problem to be solved. Indeed, if a soluble RNA polymerase moved along the DNA following the path of the double helix, the transcript should be entangled around the DNA and it is not clear how it could be untangled (for a discussion see refs. 57,58). If transcriptional complexes are fixed at the nuclear matrix, the DNA should rotate while moving through these complexes. As postulated by Liu and Wang,[59] this rotation must create twin domains of superhelical tension, a positively supercoiled domain ahead of a transcriptional complex and a negatively supercoiled domain behind it (Fig. 7.5). Positive supercoiling ahead of a transcriptional complex was proposed to facilitate transcription of DNA arranged in nucleosomes. Some data suggest that it may help to unfold the 30 nm chromatin fiber and thus to generate a chromatin conformation typical of transcriptionally active genomic regions.[60] Furthermore, according to the model of Felsenfeld and collaborators,[61] the positive supercoiling of DNA ahead of a transcriptional complex may facilitate the displacement of nucleosomes with their

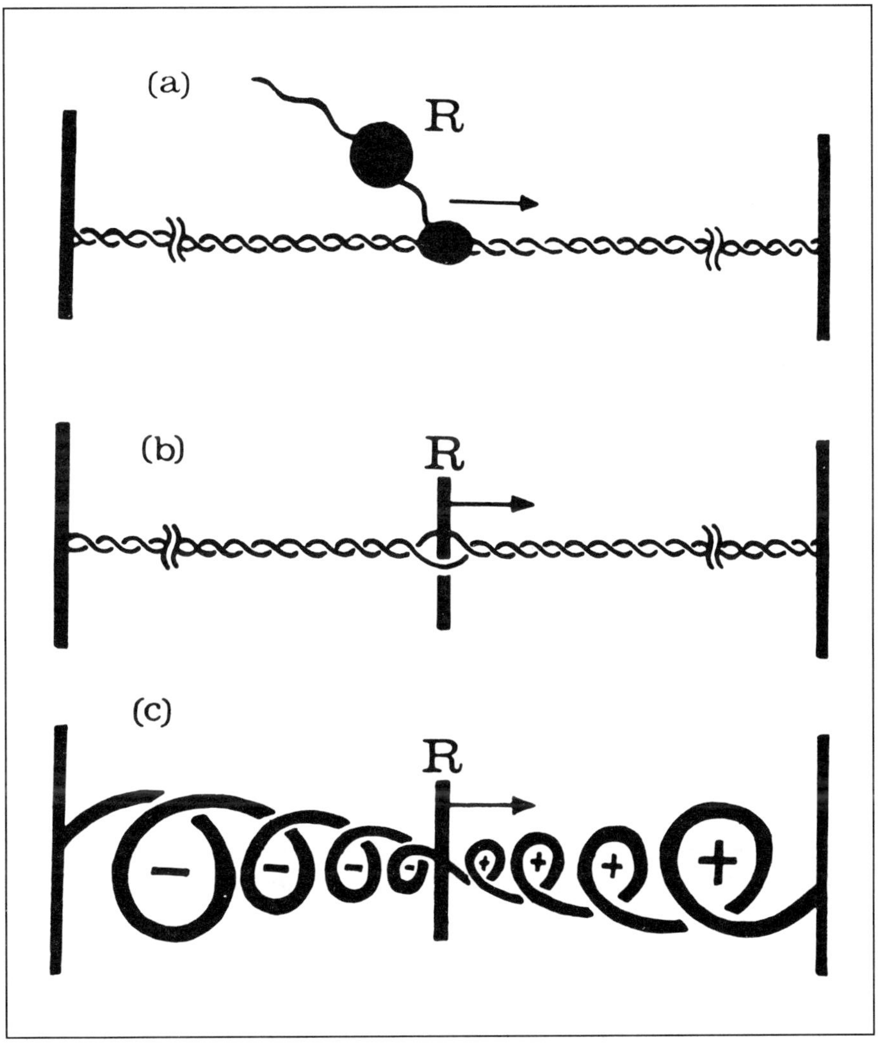

Fig. 7.5. A graphical illustration of the mechanics of transcription. (a) A transcription ensemble R including the polymerase, the nascent RNA and proteins bound to the RNA is moving in the direction of the arrow along a DNA segment; the ends of the DNA segment are anchored on a large structure represented by the solid bars. (b) The transcription ensemble can be viewed as a divider separating the helical DNA into two parts. (c) If R is moving from left to right without turning around the DNA, the DNA in front of the polymerase becomes overwound, or positively supercoiled; the DNA behind the polymerase becomes underwound, or negatively supercoiled (Note that the same will happen if DNA is moved versus immobilized transcriptional complex). (Reprinted with permission of J. Wang from Liu LF, Wang JC. Proc Natl Acad Sci USA 1987; 84:7024-7027. Copyright to individual articles retained by the authors.)

subsequent movement to the negatively supercoiled domain behind the transcriptional complex.[61] Indeed, it was found that under equilibrium conditions of histone exchange, octamers preferentially transferred from positively to negatively supercoiled DNA.[61] It is clear that temporal displacement of nucleosomes located just ahead of transcriptional complexes would highly facilitate transcription of DNA arranged in chromatin. Other data suggest that positive supercoiling of DNA can induce unfolding of nucleosomes again facilitating transcription of DNA arranged in chromatin.[62] However, the nucleosome displacement model has gained a better experimental support[63-67] (for a review see refs. 68-71). In any case, positive supercoiling of DNA, which may be generated in the course of transcription only if the rotation of RNA polymerase is restricted (Fig. 7.5 and refs. 58, 59), seems to constitute an important tool permitting the transcription "through nucleosomes" to be achieved.

The observations made with so-called "Miller spreads"[72,73] seem to be in direct contradiction with the assumption that transcriptional complexes are fixed at some internal nuclear structure. Indeed, on these spreads it is possible to see active transcriptional complexes bound to DNA, but not to any supporting structure. At a high density of transcription (a classical example is transcription of nucleolar genes by RNA polymerase I) one can see "Christmas tree-like" pictures with many transcriptional complexes moving along the transcription unit and nascent RNAs gradually increasing in size toward the end of the transcription unit. In connection with these results, it should be stressed that "Miller spreads" are prepared by dropping nuclei in a hypotonic buffer. No association of either active genes or transcriptional complexes with the nuclear matrix was observed when nuclei were treated with hypotonic solutions.[54,74] As has been discussed above (chapter 1), in hypotonic buffers the internal nuclear matrix may be partially disassembled. Hence, the structures providing a surface for the attachment of transcriptional complexes may be simply lost during the preparation of "Miller spreads."

7.3. CONCLUSIONS

The enzymes involved in replication, transcription and repair of damaged DNA are fixed at the nuclear matrix. They fulfill their function while the DNA template is locally moving along (or through) immobilized enzyme complexes. Correspondingly, any genomic fragment may be transiently bound to the nuclear matrix. The probability of transient associations of transcriptionally active sequences with the nuclear matrix is much higher because the number of simultaneously working transcriptional complexes is higher as compared to the number of replication forks and sites of DNA repair. The upstream regulatory regions of active and potentially active genes are independently bound to the nuclear matrix due to the interaction of target sequences with insoluble transcription factors and other sequence-specific DNA binding proteins. These associations are not affected by the inhibition of transcription.

Related Review Articles

Cook PR. The nucleoskeleton and the topology of transcription. Eur J Biochem 1989; 185:487-501.

Cook PR. The nucleoskeleton and topology of replication. Cell 1991; 66:627-635.

Berezney R. The nuclear matrix: a heuristic model for investigating genomic organization and function in the cell nucleus. J Cell Biochem 1991; 47:109-123.

Cook PR. RNA polymerase: structural determinant of the chromatin loop and chromosome. Bioessays 1994; 16:425-430.

Jackson DA, Cook PR. The structural basis of nuclear function. Int Rev Cytol 1995; 162A:125-149.

Stein GS, Van Wijnen AJ, Stein J et al. Contributions of nuclear architecture to transcriptional control. Int Rev Cytol 1995; 162A:251-278.

References

1. Berezney R, Coffey DS. Nuclear protein matrix: association with newly synthesized DNA. Science 1975; 189:291-292.
2. Jackson DA, McCready SJ, Cook PR. RNA is synthesized at the nuclear cage. Nature 1981; 292:552-555.
3. McCready SJ, Cook PR. Lesions induced in DNA by ultraviolet light are repaired at the nuclear cage. J Cell Sci 1984; 70:189-196.
4. Harless J, Hewitt RR. Intranuclear localization of UV-induced DNA repair in human VA13 cells. Mutat Res 1987; 183:177-184.
5. Mullenders LHF, Van Zeeland AA, Natarajan AT. Analysis of the distribution of DNA repair patches in the DNA-nuclear matrix complex from human cells. Biochim Biophys Acta 1983; 740: 428-435.
6. Mullenders LHF, van Kasteran-van Leeuwen AC, Van Zeeland A et al. Nuclear matrix associated DNA is preferentially repaired in normal human fibroblasts, exposed to a low dose of ultraviolet light

but not in Cockayne's syndrome fibroblasts. Nucl Acids Res 1988; 16:10607-10623.
7. Smith HC, Puvion E, Buchholtz LA et al. Spatial distribution of DNA loop attachment and replication sites in the nuclear matrix. J Cell Biol 1984; 99:1794-1802.
8. Fakan S, Leduk Y, Lamarre D et al. Immunoelectron microscopical distribution of poly(ADP-ribose)polymerase in the mammalian cell nucleus. Exp Cell Res 1988; 179:517-526.
9. Smith HC, Berezney R. DNA polymerase alpha is tightly bound to the nuclear matrix of actively replicating liver. Biochem Biophys Res Commun 1980; 97:1541-1547.
10. Smith HC, Berezney R. Dynamic domains of DNA polymerase alpha in regenerating rat liver. Biochemistry 1983; 22:3042-3046.
11. Foster KA, Collins JM. The interrelation between DNA synthesis rates and DNA polymerases bound to the nuclear matrix in synchronized HeLa cells. J Biol Chem 1985; 260:4229-4235.
12. Wood SH, Collins JM. Preferential binding of DNA primase to the nuclear matrix in HeLa cells. J Biol Chem 1986; 261: 7119-7122.
13. Tubo RA, Berezney R Nuclear matrix bound DNA primase. J Biol Chem 1987; 262:6637-6642.
14. Hirose F, Yamamoto S, Yamaguchi M et al. Identification and subcellular localization of the polypeptide for chick DNA primase with a specific monoclonal antibody. J Biol Chem 1988; 263:2925-2933.
15. Bravo R, Macdonald-Bravo H. Existence of two populations of cyclin/proliferating cell nuclear antigen during the cell cycle: Association with DNA replication sites. J Cell Biol 1987; 105: 1549-1554.
16. Tubo RA, Berezney R. Pre-replicative association of multiple replication enzyme activities with the nuclear matrix during rat liver regeneration. J Biol Chem 1987; 262:1148-1154.
17. Tubo RA, Berezney R. Identification of 100 and 150 S DNA polymerase alpha-primase megacomplexes solubilized from the nuclear matrix of regenerating rat liver. J Biol Chem 1987; 262:5857-5865.
18. Applegren N, Hickey RJ, Kleinschmidt AM et al. Further characterization of the human cell multiprotein DNA replication complex. J Cell Biochem 1995; 59:91-107.
19. Cook PR. The nucleoskeleton and topology of replication. Cell 1991; 66:627-635.
20. Nakayasu H, Berezney R. Mapping replication sites in eukaryotic cell nucleus. J Cell Biol 1989; 108:1-11.
21. Neri LM, Mazzotti G, Capitani S et al. Nuclear matrix-bound replication sites detected in situ by 5'-bromodeoxyuridine. Histochemistry 1992; 98:19-32.
22. Hassan AB, Cook PR. Visualization of replication sites in unfixed human cells. J Cell Sci 1993; 105:541-550.

23. Hozak P, Hassan AB, Jackson DA et al. Visualization of replication factories attached to the nucleoskeleton. Cell 1993; 73:361-373.
24. Hozak P, Jackson DA, Cook PR. Replication factories and nuclear bodies: the ultrastructural characterization of replication sites during the cell cycle. J Cell Sci 1994. 107; 2191-2202.
25. Hand R. Regulation of DNA replication on subchromosomal units of mammalian cells. J Cell Biol 1975; 64:89-97.
26. Hand R. Eukaryotic DNA: organization of the genome for replication. Cell 1978; 15:317-325.
27. Painter RB, Young BR. Formation of nascent DNA molecules during inhibition of replication initiation in mammalian cells. Biochim Biophys Acta 1976; 418:146-153.
28. Jost J-P, Seldran M. Association of transcriptionally active vitellogenin II gene with the nuclear matrix of chicken liver. EMBO J 1984; 3:2205-2208.
29. Small D, Nelkin B, Vogelstein B. The association of transcribed genes with the nuclear matrix of *Drosophila* cells during heat shock. Nucl Acids Res 1985; 13:2413-2431.
30. Iarovaia OV, Gorelova TV, Shuppe NG et al. A pattern of the heat shock genes attachment to the nuclear matrix does not depend on their functional state. Genetica (Moscow) 1985; 21: 896-901.
31. Razin SV. DNA interactions with the nuclear matrix and spatial organization of replication and transcription. Bioessays 1987; 6:19-23.
32. Razin SV, Vassetzky YS. Domain organization of eukaryotic genome. Cell Biol Int Rep 1992; 16:697-708.
33. Dworetzky SI, Wright KL, Fey EG et al. Sequence-specific DNA binding proteins are components of a nuclear matrix attachment site. Proc Natl Acad Sci USA 1992; 89:4178-4182.
34. Bidwell JP, van Wijnen AJ, Fey, EG et al. Osteocalcin gene promoter-binding factors are tissue-specific nuclear matrix components. Proc Natl Acad Sci USA 1993; 90:3162-3166.
35. Van Wijnen AJ, Bidwell JP, Fey EG et al. Nuclear matrix association of multiple sequence-specific DNA binding activities related to SP1, ATF, CCAAT, C/ERB, OCT-1 and AP-1. Biochemistry 1993; 32:8397-83402.
36. Zenk DW, Ginder GD, Brotherton TW A nuclear matrix protein binds very tightly to DNA in the avian b-globin gene enhancer. Biochemistry 1990; 29: 5221-5226.
37. Merriman HL, Van Wijnen AJ, Hiebert S et al. The tissue-specific nuclear matrix protein, NMP-2, is a member of the AML/CBF/PEBP2/runt domain transcription factor family: Interactions with the osteocalcin gene promoter. Biochemistry 1995; 34:13125-13132.
38. Nardozza TA, Quigley MM, Getzenberg RH. Association of transcription factors with the nuclear matrix. J Cell Biochem 1996; 61:467-477.

39. Reitman M, Lee E, Westphal H et al. An enhancer/locus control region is not sufficient to open chromatin. Mol Cell Biol 1993; 13:3990-3998.
40. Ptashne M. How eukaryotic transcriptional activators work. Nature 1988; 335:683-689.
41. Krajewski WA, Razin SV. DNA-protein interactions and spatial organization of DNA. Mol Biol Rep 1993; 18:167-175.
42. Boulikas T. How enhancers work: Juxtapositioning of DNA control elements by synergistic interaction of MARs. Int J Oncol 1995; 6:1313-1318.
43. Morrison NA, Shine J, Fragonas JC et al. 1,25-dihydroxyvitamin D-responsive element and glucocorticoid repression in osteocalcin gene. Science 1989; 246:1158-1161.
44. Stromstedt PE, Poellinger L, Gustafsson JA et al. The glucocorticoid receptor binds to a sequence overlapping the TATA box of the human osteocalcin promoter: A potential mechanism of negative regulation. Mol Cell Biol 1991; 11:3379-3383.
45. Kerner SA, Scott RA, Pike JW. Sequence elements in human osteocalcin gene confer basal activation and inducible response to hormonal vitamin D_3. Proc Natl Acad Sci USA 1989; 86: 4455-4459.
46. Heinrichs AAJ, Bortell L, Rahman S et al. Identification of multiple glucocorticoid receptor binding sites in the rat osteocalcin gene promoter. Biochemistry 1993; 32:11436-11444.
47. Truss M, Beato M. Steroid hormone receptors: Interaction with deoxyribonucleic acid and transcription factors. Endocr Rev 1993; 14:459-479.
48. Glass CK. Differential recognition of target genes by nuclear receptor monomers, dimers and heterodimers. Endocr Rev 1994; 15:391-407.
49. Ruh MF, Dunn R, Ruh TS. Interrelationships between nuclear structure and ligand-activated intracellular receptors. Crit Rev Eukar Gene Express 1996; 6:271-283.
50. Xing Y, Johnson CV, Moen Jr PT et al. Nonrandom gene organization: Structural arrangements of specific pre-mRNA transcription and splicing with SC-35 domains. J Cell Biol 1995; 131:1635-1647.
51. Andreeva M, Markova D, Loidl P et al. Intranuclear compartmentalization of transcribed and nontranscribed c-myc sequences in Namalva-S cells. Eur J Biochem 1992; 207:887-894.
52. Jackson DA, McCready SJ, Cook PR. Replication and transcription depend on attachment of DNA to the nuclear cage. J Cell Sci 1984; Suppl. 1:59-79.
53. Razin SV, Yarovaya OV. Initiated complexes of RNA polymerase II are concentrated in the nuclear skeleton associated DNA. Exp Cell Res 1985; 158:273-275.
54. Jackson DA, Cook PR. Transcription occurs at a nucleoskeleton. EMBO J 1985; 4:919-925.

55. Diskinson P, Cook PR, Jackson DA. Active RNA polymerase I is fixed within the nucleus of HeLa cells. EMBO J 1990; 9:2207-2214.
56. Jackson DA, Cook PR. Transcriptionally-active minichromosomes are attached transiently in nuclei through transcription units. J Cell Sci 1993; 105: 1143-1150.
57. Cook PR, Gove F. Transcription by an immobilized RNA polymerase from bacteriophage T7 and the topology of transcription. Nucl Acids Res 1992; 20:3591-3598.
58. Cook PR. The nucleoskeleton and topology of transcription. Eur J Biochem 1989; 185:487-501.
59. Liu LF, Wang JC. Supercoiling of the DNA template during transcription. Proc Natl Acad Sci USA 1987; 84:7024-7027.
60. Lee M-S, Garrard WT. Positive DNA supercoiling generates a chromatin conformation characteristic of highly active genes. Proc Natl Acad Sci USA 1991; 88:9675-9679.
61. Clark DJ, Felsenfeld G. Formation of nucleosomes on positively supercoiled DNA. EMBO J 1991; 10:387-395.
62. Lee M-S, Garrard WT. Transcription-induced nucleosome "splitting": an underlying structure for DNase I sensitive chromatin. EMBO J 1991; 10:607-615.
63. Pfaffle P, Gerlach V, Bunzel L et al. In vivo evidence that transcription-induced stress causes nucleosome dissolution and regeneration. J Biol Chem 1990; 265:16830-16840.
64. Clark DJ, Felsenfeld G. A nucleosome core is transferred out of the path of a transcribing polymerase. Cell 1992; 71:11-22.
65. O'Donnohue M-F, Duband-Goulet I, Hamiche A et al. Octamer displacement and redistribution in transcription of single nucleosomes. Nucl Acids Res 1994; 22:937-945.
66. Studitsky VM, Clark DJ, Felsenfeld G. A histone octamer can step around a transcription polymerase without leaving the template. Cell 1994; 76:371-382.
67. Studitsky VM, Clark DJ, Felsenfeld G. Overcoming a nucleosomal barrier to transcription. Cell 1995; 83:19-27.
68. Kornberg RD, Lorch Y. Irresistible force meets immovable object: transcription and the nucleosome. Cell 1991; 67:833-836.
69. Thoma F. Structural changes in nucleosomes during transcription: strip, split or flip? Trends Genet 1991; 7:175-177.
70. Morse RH. Transcribed chromatin. Trends Biochem Sci 1992; 17:23-26.
71. Wolffe A. Nucleosome positioning and modification: Chromatin structures that potentiate transcription. Trends Biochem Sci 1994; 19:240-244.
72. Miller OL. The nucleolus, chromosomes and visualization of genetic activity. J Cell Biol 1981; 91:15s-27s.
73. Miller OL. Some ultrastructural aspects of genetic activity in eukaryotes. J Cell Sci 1984; Suppl. 1:81-93.

74. Razin SV, Yarovaia OV, Georgiev GP. Low ionic strength extraction of nuclease-treated nuclei destroys the attachment of transcriptionally active DNA to the nuclear skeleton. Nucl Acids Res 1985; 13:7427-7444.

CHAPTER 8

DNA Loop Anchorage Sites Constitute Preferential Targets for DNA Cleavage by Nucleases

8.1. LARGE-SCALE FRAGMENTATION OF GENOMIC DNA AT THE INITIAL STEPS OF APOPTOSIS PROCEEDS VIA EXCISION OF CHROMOSOMAL DNA LOOPS AND THEIR OLIGOMERS

Recent evidence suggests that the degradation of genomic DNA in the course of apoptosis starts with an excision of 50-300 kb DNA fragments[1-7] which are then further digested to oligonucleosomal ladders. However, in some cases no oligonucleosomal ladders were observed at all while loop size DNA fragments appeared as usual.[3] It was proposed that the pattern of large-scale fragmentation of chromosomal DNA reflected the periodicity of 30 nm chromatin fibril organization into loops attached to the nuclear matrix.[1,2,8,9] The only reason for this proposition was however a certain similarity in sizes. Although the specificity of long-range DNA fragmentation by nucleases within any characterized genomic area can easily be analyzed, the pattern of this fragmentation could not be directly compared with the partitioning of DNA into loops because of the absence of a reliable procedure for mapping DNA loop anchorage regions (LARs). However, this kind of comparison became possible with the elaboration of an approach for chromosomal DNA loop excision by topoisomerase II-mediated DNA cleavage at matrix attachment sites (discussed in chapter 5). Using this approach, Razin and collaborators have compared the patterns of long-range fragmentation of nucleolar and euchromatic genes in mouse and human cells triggered to undergo apoptosis either by the tumor necrosis factor or by serum deprivation with the pattern generated by high-salt-insoluble topoisomerase II.[10] The patterns were found to be essentially similar if not identical (Fig. 8.1). A conclusion has been drawn that the large-scale fragmentation of eukaryotic DNA at the initial steps of apoptosis proceeds in a specific fashion via excision of chromosomal DNA loops and their oligomers.[10]

In contrast to most of the other authors,[1-3] Razin and collaborators failed to observe a discontinuity in a size distribution of large DNA fragments excised at the initial steps of apoptosis (Fig. 8.1A), unless the attention was focused on a particular genomic area by hybridization with specific probes (Fig. 8.1B). The reason for this discrepancy is not clear at the moment. One cannot exclude that the 300 kb band observed previously by several authors[1-3] is a compression artefact of the field inversion gel electrophoresis used for separation of the excised DNA fragments. Indeed, the sharpness of the above band varied significantly in the experiments of different authors (one may for example compare the data presented in refs. 1 and 2). Furthermore, in no case the

Fig. 8.1 (opposite). Analysis of the specificity of nucleolar genes long-range cleavage in cells undergoing apoptosis triggered by TNF-α. (A) Separation of the released DNA fragments by PFGE (staining with ethidium bromide). (B) Hybridization of the separated DNA fragments (after Southern transfer) with the ^{32}P-labeled rDNA probe (autoradiograph). (lane 1) Fragmentation of DNA by the high-salt-insoluble topoisomerase II of the nuclear matrix (cleavage of DNA at the basements of the loops). (lanes 2-4) Fragmentation of DNA in RPMI-6410t cells pretreated with actinomycin D (1 μg/ml) and cultivated for 48 h in the absence of TNF-α (lane 2), in the presence of 5 Un/ml of TNF-α (lane 3) and in the presence of 10 Un /ml of TNF-α (lane 4). (lanes 5-7) fragmentation of DNA in mouse L929 cells cultivated for 0 h (lane 5), 12 h (lane 6) and 18 h (lane 7) in the medium containing 5 Un/ml of TNF-α. All lanes were loaded with DNA from the same amount of cells. The start area (i.e., where the agarose blocks with embedded DNA were inserted in the gel) is not shown. Arrows at the left side of the panel A indicate positions of the molecular weight markers. (Reprinted with permission of The American Society for Biochemistry and Molecular Biology from Lagarkova MA, Iarovaia OV, Razin SV. J Biol Chem 1995; 270:20239-20241. © 1995 by The American Society for Biochemistry and Molecular Biology.)

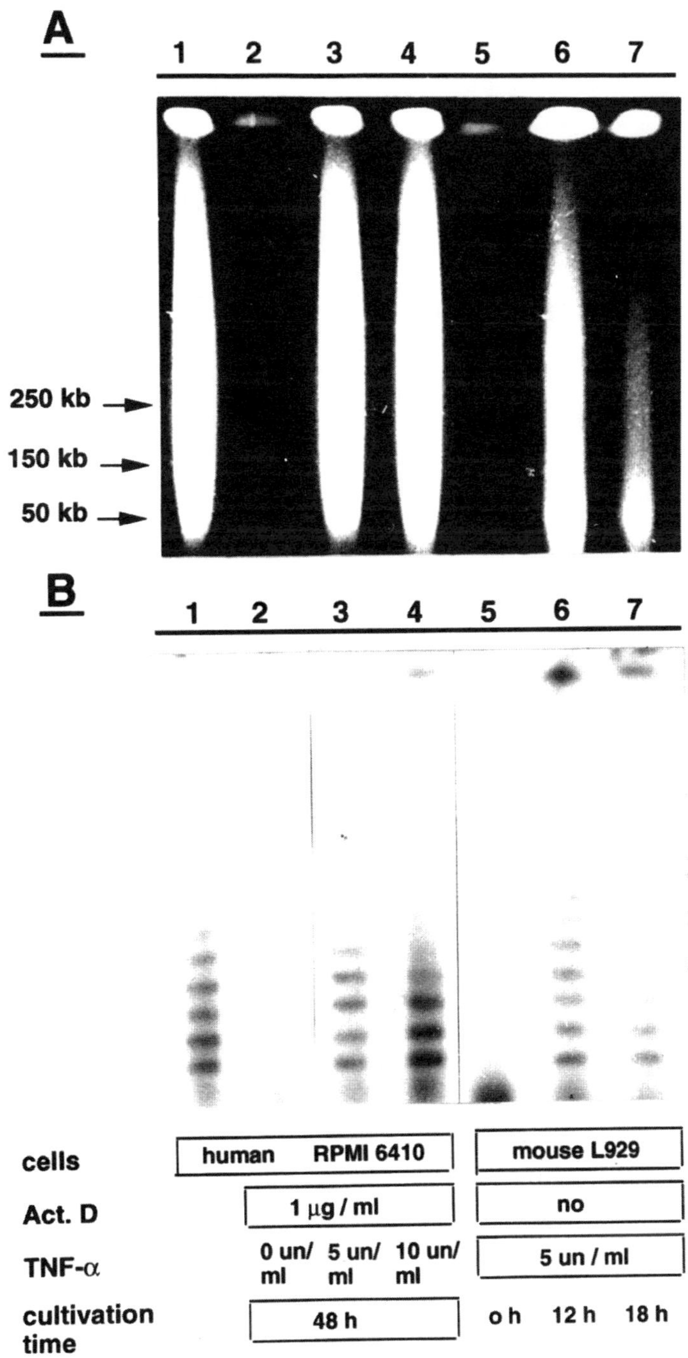
A
1 2 3 4 5 6 7
250 kb
150 kb
50 kb
B
1 2 3 4 5 6 7
cells
human RPMI 6410
mouse L929
Act. D
1 μg / ml
no
TNF-α
0 un/ ml
5 un/ ml
10 un/ ml
5 un / ml
cultivation time
48 h
o h
12 h
18 h

above cited authors showed proportional separation of the phage λ DNA concatemers within the size area ranging from 50-300 kb (i.e., where the discontinuous distribution of genomic DNA fragments was observed). Finally, when other researchers used PFGE for the separation of the excised large DNA fragments, no discontinuity in the size distribution of these fragments was observed.[6] In the experiment shown in Figure 8.1, the compression artefact was observed close to the top of the gel (the relatively sharp band of about 700 kb). This compression band accumulated fragments ranging in sizes from 700 kb to more then 1000 kb.

The excision of chromosomal DNA loops at the initial steps of apoptosis may be explained in two different ways. The first explanation implies some nuclease which is deposited at the nuclear matrix and is normally inactive. Activation of this enzyme may then constitute an important step in the cascade of events resulting in apoptotic cell death. This explanation is indirectly corroborated by the finding that a nuclease cleaving DNA into loop-sized fragments at the initial steps of apoptosis is distinct from the common nuclear Ca^{2+}/Mg^{2+} dependent nuclease because the former nuclease is not inhibited by Zn^{2+} ions[2] although Zn^{2+} is a powerful inhibitor of the latter one.[11] Although apoptosis can be triggered by an exposure of cells to topoisomerase II-specific drugs, it is not likely that topoisomerase II is directly involved in the cleavage of genomic DNA into large fragments at the initial steps of apoptosis. Indeed, similar patterns of DNA cleavage into 50 kb fragments were observed in thymocytes triggered to undergo apoptosis by topoisomerase II-specific drugs and by dexametasone. Furthermore, in both cases this fragmentation was suppressed by actinomycin D suggesting that the cleavage of genomic DNA into 50 kb fragments is a complex process requiring the expression of new genes[12] (for a discussion see refs. 8, 9).

The excision of DNA loops and their oligomers at the initial steps of apoptosis may also be explained by supposing that for some reasons DNA loop anchorage sites in living cells are preferentially accessible to nucleases. In this case, they can be selectively digested by a variety of nucleases (including the cytoplasmic ones) which may get access to the nuclei at the beginning of apoptosis. The data discussed in the next section strongly support this supposition.

8.2. BASEMENTS OF CHROMOSOMAL DNA LOOPS CONSTITUTE PREFERENTIAL TARGETS FOR EXOGENOUS AND ENDOGENOUS NUCLEASES

In order to find out whether the basements of chromosomal DNA loops are preferential targets for nucleases, the patterns of the nuclear DNA long-range fragmentation by endogenous and exogenous nucleases have been studied.[13,14] Permeabilized cells were treated with exogenous nucleases or incubated under conditions favoring autodigestion of DNA by endogenous nucleases. This was followed by pulsed-field-gel-electophoretic separation of cleavage products and by Southern analysis of the specificity of fragmentation of the *c-myc* amplicone and the nucleolar genes.[13,14] Again, the patterns of long-range DNA fragmentation by nucleases were compared with the pattern of DNA cleavage at loop basements by high-salt-insoluble topoisomerase II (Fig. 8.2). Two important observations have been made in these experiments:

First, it has been demonstrated that the loop anchorage regions are preferentially sensitive to a variety of nucleases including DNase I. The preferential sensitivity to DNase I was lost after disruption of the chromatin structure by 2M NaCl extraction.[13] Hence, the loop anchorage regions contain strong DNase I hypersensitive sites (or clusters of sites) determined at the level of DNA packaging in chromatin. Among endogenous nucleases, the targeting for loop anchorage regions was most clearly observed for those possessing the properties of DNase II (acid endonuclease active in the presence of either EDTA or Zn^{2+} ions which both are known to inhibit the nuclear Ca^{2+}/Mg^{2+}-dependent endonuclease). In connection with the above observations, it is worth reminding that the cleavage of chromosomal DNA into loop-sized fragments at the initial stages of apoptosis also is not inhibited by Zn^{2+} ions,[2] while the presence of these ions in a cell culture medium strongly suppresses the generation of oligonucleosomal ladders.[15-17]

Second, it has been found that the loop anchorage regions are preferentially accessible to cleavage by single-stranded DNA-specific nucleases (S1 and mung bean nuclease) and by nuclease Bal 31.[13,14,18] It is important to note that the above results could not be reproduced when purified DNA was treated with single-stranded DNA-specific nucleases either under standard conditions or under conditions favoring partial denaturation of DNA.[13] Hence,

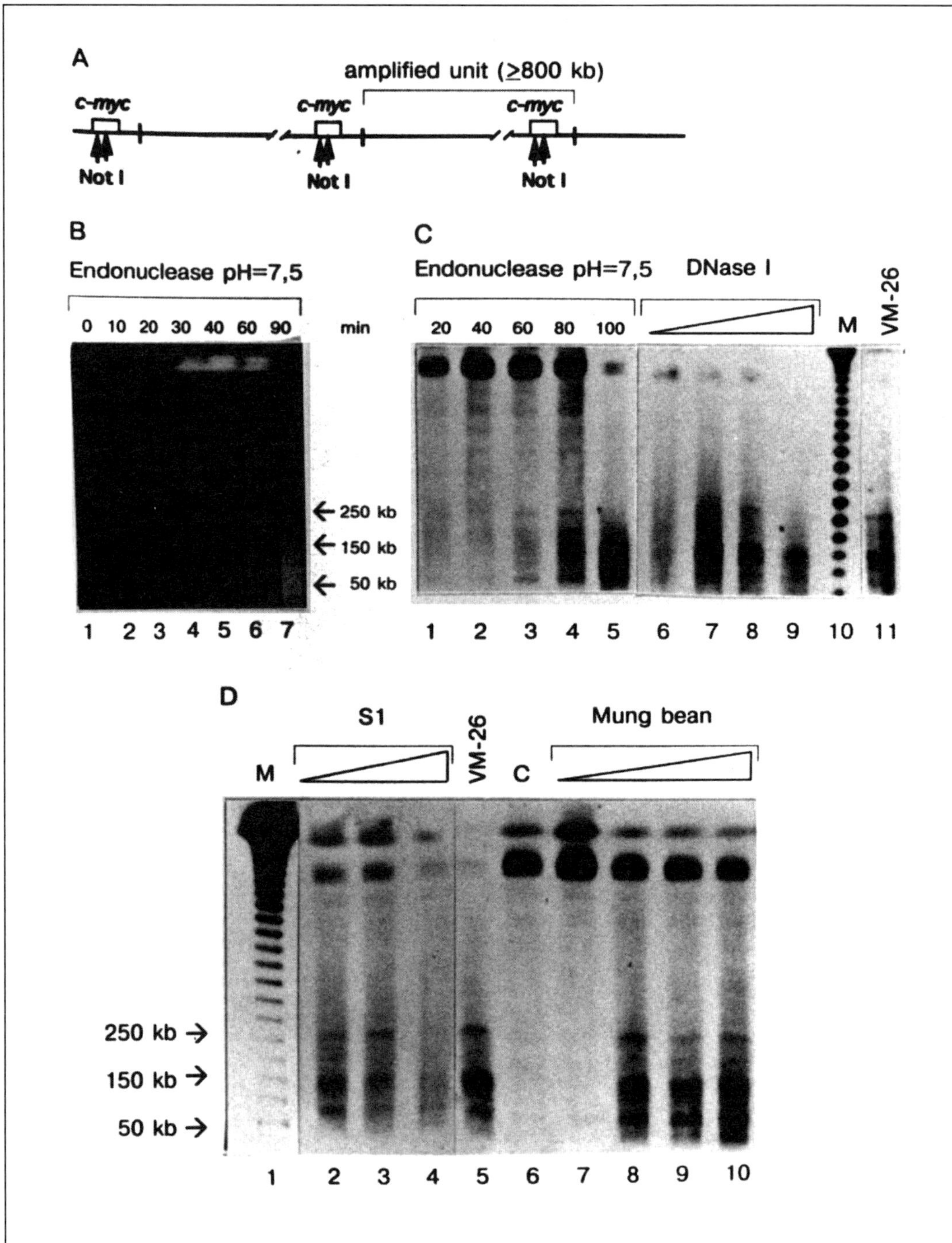

A
amplified unit (≥800 kb)
c-myc
c-myc
c-myc
Not I
Not I
Not I
B
Endonuclease pH=7,5
0 10 20 30 40 60 90 min
← 250 kb
← 150 kb
← 50 kb
1 2 3 4 5 6 7
C
Endonuclease pH=7,5
DNase I
20 40 60 80 100
M
VM-26
1 2 3 4 5 6 7 8 9 10 11
D
S1
Mung bean
M
VM-26
C
250 kb →
150 kb →
50 kb →
1 2 3 4 5 6 7 8 9 10

Fig. 8.2 (opposite). Analysis of specificity of the human c-myc *amplicone long-range fragmentation by endogenous and exogenous nucleases. (A) A map of the organization of the* c-myc *amplicone in human OC-NYH-VM cells (see ref. 8 for the details). The positions of the gene are shown by open rectangles. The third exon of the* c-myc *gene which was used as a probe in hybridization experiments is located just downstream of the pair of Not I cleavage sites indicated by arrows. (B) Size distribution of DNA fragments released from the genome of permeabilized OC-NYH-VM cells incubated for different time intervals (from 0-90 min as indicated above the slots) at 37°C in a neutral buffer supplemented with Mg^{2+} and Ca^{2+} ions (digestion with endogenous Ca^{2+}/ Mg^{2+}-dependent nucleases). After separation by PFGE the DNA fragments were stained with ethidium bromide. (C) Permeabilized cells were either incubated for different time intervals (from 0-100 min, as indicated above the slots) at 37°C in a neutral buffer favoring the digestion of DNA with endogenous nucleases (lanes 1-5) or were treated for 30 min at 0°C with different amounts of exogenous DNase I in a Mg^{2+} containing neutral buffer. The DNA samples from cells treated with 2, 10, 25 and 50 µg/ml of DNase I are shown, respectively, in lanes 6-9. After termination of incubations and digestion of proteins, all DNA samples were additionally digested with Not I (to permit indirect end-labeling). Thereafter they were separated by PFGE and subjected to Southern analysis using ^{32}P-labeled third exon of the* c-myc *gene as the hybridization probe. Lane 11 demonstrates the distribution of DNA products of topoisomerase II-mediated DNA cleavage (excised DNA loops). The distribution of λ phage DNA concatemers is shown in lane 10. (D) Permeabilized cells were incubated for 10 min at 37°C in Zn^{2+}-containing buffer (pH 5.0) supplemented with 300, 150 or 50 units, respectively of S1 nuclease (lanes 2-4) or with 60, 150 or 300 units, respectively of mung bean nuclease (lanes 8-10). After additional digestion with Not I, separation by PFGE and Southern transfer, fragments bearing the third exon of* c-myc *gene were visualized by hybridization. Lanes 6 and 7 represent DNA from nontreated cells (lane 6) and cells incubated for 10 min at 37°C in Zn^{2+}-containing buffer (pH 5.0) without enzymes. Lane 5 demonstrates the distribution of DNA products of topoisomerase II-mediated DNA cleavage (excised DNA loops). The distribution of λ phage DNA concatemers is shown in lane 1. Note that either endogenous digestion or treatment of permeabilized cells with exogenous nucleases (including the single-stranded DNA-specific nucleases) resulted in accumulation of characteristic 80, 150 and 260 kb fragments of the* c-myc *amplicone similar (if not identical) to the fragments excised from this amplicone by the topoisomerase II-mediated cleavage. (Reprinted with permission of Academic Press, Inc. from Razin SV, Gromova II, Iarovaia OV. Int Rev Cytol 1995; 162B:405-448. © 1995 Academic Press, Inc.)*

it is likely that abnormalities in the DNA double helix structure at the loop anchorage sites (i.e., in regions which can be recognized as preferential targets by S1, mung bean and Bal 31 nucleases) arise from DNA interaction with the nuclear matrix proteins. This conclusion is directly supported by the results of in vivo footprinting of the sites of DNA-protein interaction within a loop anchorage region in the upstream area of the chicken α globin genes domain.[18] Multiple regions of hypersensitivity to S1 nuclease were found also in the upstream area of the human *c-myc* gene[19] within the loop anchorage region mapped by the topoisomerase II-mediated DNA loop excision (see chapter 5). The results of other researchers further support the model postulating that loop anchorage regions may contain either permanently melted DNA sequences or DNA sequences with noncanonical secondary structures bound to specific affinity sites on the nuclear matrix.[20-24] As discussed in chapter 6, the ability to melt under superhelical tension is an intrinsic property of MAR elements.[20-22] Due to this property MAR elements are selectively recognized by several proteins, such as SATB 1 or nucleolin (see chapter 6). As reported previously, brief treatment of isolated metaphase chromosomes with Bal 31 nuclease results in the cleavage of genomic DNA into loop-sized fragments.[25] Observations described in this chapter suggest that the above result directly reflects the mode of chromosomal DNA organization into loops. Recent evidence indicates that the basements of chromosomal DNA loops are preferentially accessible to cleavage by S1 nuclease in yeast chromosomes also.[26] Possible reasons for preferential accessibility of DNA loop anchorage sites to nucleases will be discussed in the chapter 12.

It is interesting that preferential cleavage of DNA by the nuclease Bal 31 within loop anchorage regions can be observed only in proliferating cells. In permeabilized normal human lymphocytes no specific long-range fragmentation of nucleolar DNA by the nuclease Bal 31 was detected. The specific pattern of this DNA fragmentation into chromosomal loops was, however, observed after stimulation of lymphocytes to proliferation by treatment with phytohemagglutinin.[27] This may indicate that abnormalities in the DNA secondary structure which are recognized as targets by the nuclease Bal 31 (not necessarily melted regions, but also cruciforms and other noncanonical structures) reflect an active status of replication origins located within the loop anchorage regions (see the next chapter). The supposition is corroborated by recent observations of Ward and collaborators.[24,28] Using quantitative fluorescence flow cytometry (with antibodies recognizing cruciforms in DNA), these authors have found that DNA of an S-phase nucleus contains approximately (3-5) x 10^5 cruciforms. However, no cruciforms were detected in G_0 and G2/M nuclei. As has been discussed before (chapter 6 and refs. 29-33), DNA fragments isolated from the loop anchorage regions frequently contain imperfect and perfect inverted repeats which are able to form cruciforms under superhelical strain and possibly under some other conditions. It is feasible that formation of the above cruciforms constitutes an important step in activation of replication origins which are localized in the loop anchorage regions (see the next chapter). A special class of proteins binding to the inverted repeats in DNA have been described recently.[34,35] HMG 1, a MAR-binding protein which is likely to fulfill multiple different functions in eukaryotic cell nuclei, also possess an ability to recognize and bind specifically the cruciform DNA.[36] These proteins possibly stabilize the DNA cruciforms upon their formation within active replication origins.

8.3. CONCLUSIONS

Although some parts of the DNA loop anchorage regions may be protected from nucleases by the interaction with the nuclear matrix proteins, the anchorage regions as a whole are preferentially accessible to a variety of endogenous and exogenous nucleases. Some parts of anchorage regions seem to be either permanently melted or organized into a noncanonical secondary structure constituting targets for single-stranded DNA-specific nucleases. These targets are not simply easily melting DNA sequences because they cannot be recognized by the above nucleases in purified DNA under conditions favoring partial denaturation of easily-melting DNA sequences.

The observations described in this chapter suggest that DNA loops can be excised from the genome as a result of brief treatment of permeabilized cells with exogenous nucleases. DNA loops and their oligomers are also excised by endogenous nucleases at the initial steps of apoptosis.

REFERENCES

1. Filipski J, Leblank J, Youdale T et al. Periodicity of DNA folding in higher order chromatin structures. EMBO J 1990; 9:1319-1327.
2. Brown DG, Sun X-M, Cohen GM. Dexamethasone-induced apoptosis involves cleavage of DNA to large fragments prior to internucleosomal fragmentation. J Biol Chem 1993; 268:3037-3039.
3. Oberhammer F, Wilson JW, Dive C et al. Apoptotic death in epithelial cells: cleavage of DNA to 300 and/or 50 kb fragments prior to or in the absence of internucleosomal fragmentation. EMBO J 1993; 12:3679-3684.
4. Cain K, Inayat-Hussain SH, Wolfe JT et al. DNA fragmentation into 200-250 and/or 30-50 kilobase pair fragments in rat

liver nuclei is stimulated by Mg^{2+} alone and Ca^{2+}/Mg^{2+} but not by Ca^{2+} alone. FEBS Let 1994; 349:385-391.
5. Sun X-M, Cohen GM. Mg(2+) dependent cleavage of DNA into kilobase pair fragments is responsible for initial degradation of DNA in apoptosis. J Biol Chem 1994; 269:14857-14860.
6. Zhivotovsky B, Cedervall B, Jiang S et al. Involvement of Ca^{2+} in the formation of the high molecular weight DNA fragments in thymocyte apoptosis. Biochem Biophys Res Commun 1994; 202:120-127.
7. Walker PR, Kokileva L, Leblank J et al. Detection of the initial stages of DNA fragmentation in apoptosis. Biotechniques 1993; 15:1032-1041.
8. Walker PR, Sikorska M. Endonuclease activities, chromatin structure, and DNA degradation in apoptosis. Biochem Cell Biol 1994; 72:615-623.
9. Walker PR, Pandey S, Sikorska M. Degradation of chromatin in apoptotic cells. Cell Death Differ 1995; 2:97-104.
10. Lagarkova MA, Iarovaia OV, Razin SV. The large-scale fragmentation of mammalian DNA in the course of apoptosis proceeds via excision of chromosomal DNA loops and their oligomers. J Biol Chem 1995; 270:20239-20241.
11. Pietsch MC, Mannherz HG, Tschopp J. The apoptosis endonucleases: cleaning up after cell death? Trends in Cell Biol 1994; 4:37-41.
12. Walker PR, Smith C, Youdale T et al. Topoisomerase II-reactive chemoterapeutic drugs induce apoptosis in thymocytes. Cancer Res 1991; 51:1078-1085.
13. Gromova II, Nielsen OF, Razin SV. Long-range fragmentation of the eukaryotic genome by exogenous and endogenous nucleases proceeds in a specific fashion via preferential DNA cleavage at matrix attachment sites. J Biol Chem 1995; 270:18685-18690.
14. Lagarkova MA, Iarovaia OV, Razin SV. Excision of chromosomal DNA loops by pretreatment of permeabilised cells with Bal 31 nuclease. Mol Gen Genet 1995; 249:253-256.
15. Cohen JJ, Duke RC. Glucocorticoid activation of a calcium-dependent endonuclease in thymocyte nuclei leads to cell death. J Immunol 1984; 132:38-42.
16. Arends MJ, Morris RG, Wyllie AH. Apoptosis. The role of the endonuclease. Am J Pathol 1990; 136:593-608.
17. Peitsch M, Polzar B, Stephan H et al. Characterization of the endogenous deoxyribonuclease involved in nuclear DNA degradation during apoptosis (programmed cell death). EMBO J 1993; 12:371-377.
18. Recillas-Targa F, Razin S, De Moura Gallo CV et al. Excision close to matrix attachment regions of the entire chicken α-globin genes domain by nuclease S1 and characterization of framing structures. Proc Natl Acad Sci USA 1994; 91:4422-4426.

19. Michelotti GA, Michelotti EF, Pullner A et al. Multiple single-stranded cis elements are associated with activated chromatin of the human *c-myc* gene in vivo. Mol Cell Biol 1996; 6:2656-2669.
20. Kohwi-Shigematsu T, Kohwi Y. Torsional stress extended base unpairing in suppressor sites flanking immunoglobulin heavy chain enhancer. Biochemistry 1990; 29:9551-9560.
21. Bode J, Kohwi I, Dickinson L et al. Biological significance of unwinding capability of nuclear matrix-associated DNAs. Science 1992; 255:195-197.
22. Bode J, Schlake T, Rios-Ramirez M et al. Scaffold/matrix-attached regions: Structural properties creating transcriptionally active loci. Int Rev Cytol 1995; 162A:389-454.
23. Paul A-L, Ferl RJ. Osmium tetroxide footprinting of a scaffold attachment region in the maize Adh 1 promoter. Plant Mol Biol 22:1145-1151.
24. Ward GK, Shihab-el-Deen A, Zannis-Hadjopoulos M et al. DNA cruciforms and the nuclear supporting structure. Exp Cell Res 1991; 195:92-98.
25. Clark RW, Tseng PO, Lechuga JM. A nuclease-derived fragments of metaphase DNA and its relationship to the replicon. Exp Cell Res 1987; 169:296-310.
26. Svetlova EYu, Razin SV, Filipski J. Mapping of the nuclease S1-hypersensitive sites in chromosome III of *Saccharomyces cerevisiae.* Dokladi AN (in Russian) 1996; (in press).
27. Iarovaia OV, Lagarkova MA, Razin SV. The specificity of human lymphocyte nucleolar DNA long-range fragmentation by endogenous topoisomerase II and exogenous Bal 31 nuclease depends on cell proliferation status. Biochemistry 1995; 43:4133-4138.
28. Ward GK, McKenize R, Zannis-Hadjopoulos M et al. The dynamic distribution and quantitation of DNA cruciforms in eukaryotic nuclei. Exp Cell Res 1990; 188:235-246.
29. Kalandadze AG, Razin SV, Georgiev GP. Cloning and analysis of DNA fragments permanently attached to the nuclear matrix. Doklady AN SSSR (Moscow) 1988; 300:1001-1005.
30. Goldberg GI, Collier I, Cassel A. Specific DNA sequences associated with the nuclear matrix in synchronized mouse 3T3 cells. Proc Natl Acad Sci USA 1983; 80:6887-6891.
31. Opstelten RJG, Clement JME, Wanka F. Direct repeats at nuclear matrix-associated DNA regions and their putative control function in the replicating eukaryotic genome. Chromosoma 1989; 98:422-427.
32. Boulikas T, Kong CF. Multitude of inverted repeats characterizes a class of anchorage sites of chromatin loops to the nuclear matrix. J Cell Biochem 1993; 53:1-12.

33. Boulikas T, Kong CF. A novel class of matrix attached regions (MARs) identified by random cloning and their implication in differentiation and carcinogenesis. Int J Oncol 1993; 2:325-330.
34. Pearson CE, Ruiz MT, Price GB et al. Cruciform DNA-binding protein in HeLa cell extracts. Biochemistry 1994; 33:14185-14196.
35. Pearson CE, Zannis-Hadjopoulos M, Price GB et al. A novel type of interaction between cruciform DNA and cruciform binding protein from HeLa cells. EMBO J 1995; 14:1571-1580.
36. Bianchi ME, Beltrame M, Paonessa G. Specific recognition of cruciform DNA by nuclear protein MHG1. Science 1989; 243: 1056-1059.

CHAPTER 9

Replication Origins Are Located at or Close to DNA Loop Anchorage Sites

9.1. SEVERAL WORDS ABOUT REPLICATION ORIGINS IN HIGHER EUKARYOTES

The genome of higher eukaryotes is subdivided into a number of quasi-independent replication units.[1] They are separated by replication termination sites, and each unit seems to have a specific place where the replication starts (*ori*). Great advances have recently been made in characterization of replication origins in yeast (for a review see refs. 2-5). These advances became possible due to the discovery of DNA sequences able to support autonomous replication of plasmids in yeast cells (ARS elements).[6] Unfortunately, numerous attempts to apply the same technique to obtain autonomous replication of plasmids in mammalian cells yielded very controversial and often irreproducible results,[7-10] except for the transfection of plasmids containing papovavirus *ori* into T antigen-producing cells.[11,12] Present knowledge about the mechanisms of replication initiation and progression in higher eukaryotes is partially based on the results obtained in experiments with replication of minichromosomes of DNA-containing viruses (see ref. 5 for a review). However, it is not clear whether these minichromosomes represent a good model of eukaryotic replicons.

A variety of procedures for mapping replication origins in cloned areas of mammalian genomes have been developed recently (for a review see refs. 2, 5, 13-15). Unfortunately, the application of different techniques for mapping replication origins in the same genomic area within the DHFR amplicone of Chinese hamster cells has resulted in controversial conclusions. While some results apparently indicate that the replication origin can be mapped to a high accuracy (at best case within a 400 bp DNA fragment),[16] the application of another experimental approach to the same genomic area justifies the conclusion that replication may start within a large region (more than 20 kb) with almost equal probability.[17] Nevertheless, all the results support the idea that replication initiation occurs nonrandomly in the genome of living cells. This conclusion is corroborated by previous observations demonstrating that the time of replication of a gene depends on the position of this gene in the genome.[18] Transfer of a gene to other positions in the genome changes the timing of its replication.[19] On the other hand, studies of replication of circular plasmid or viral DNA injected into *Xenopus laevis* eggs revealed no specificity of replication initiation.[20-22] Hence, in some cases the initiation of replication may occur at any position. A specific initiation at the origins mapped in living cells was, however, observed when isolated nuclei were placed in an extract of *Xenopus* eggs.[23] Furthermore, initiation of replication on exogenous DNA placed in an extract of *Xenopus* eggs strictly depends on assembly of a functional nucleus.[24,25] These observations directly demonstrate the importance of a proper organization of DNA within nuclei for correct initiation of replication.

9.2. PERMANENT ATTACHMENT OF REPLICATION ORIGINS TO THE NUCLEAR MATRIX

Initial observations related to the problem of spatial organization of replication units in eukaryotic cell nuclei were made by Buongiorno-Nardelli and collaborators.[26] It was found that the size of supercoiled loop domains in the genomes of different eukaryotic cells was directly related to the average size of replicons in these cells (Fig. 9.1). Similar observations were made by Hartwig.[27] These observations actually suggested that individual loops corresponded to replicons. One may wonder how the above relation can be revealed in active nuclei where transcribed genes are attached to the nuclear matrix. However, the method used by Buongiorno-Nardelli and collaborators for determining an average loop size (by measuring the diameter of the relaxed DNA loop

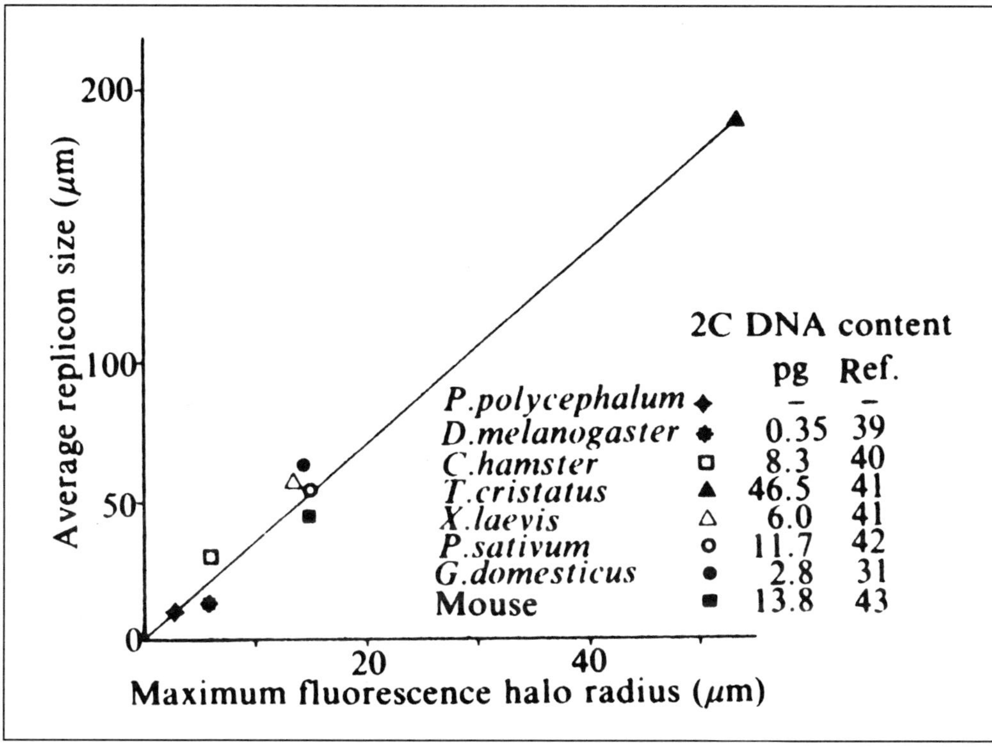

Fig. 9.1. Correlation between average replicon sizes and average sizes of chromosomal DNA loops (determined by measurement of the DNA loop halo radius) in cells of different organisms. Reprinted with permission of Macmillan Magazines Limited from Buongiorno-Nardelli M, Gioacchino M, Carri MT et al. Nature 1982; 298:100-102. © 1982 Macmillan Magazines Limited.)

halo in nucleoids) made it possible to estimate an average size of only large loops (i.e., loops which are not divided into sub-domains by the attachment of transcribing DNA sequences to the nuclear matrix; see discussion in chapter 3).

Although the correspondence of DNA loops to replicons suggests that some specific element of each replicon (and there are only two such elements, a replication origin and a termination site) is attached to the nuclear matrix, it does not show which of the above two elements is actually bound to the nuclear matrix. The first observations indicating that replication origins are attached to the nuclear matrix were made by Wanka and collaborators in pulse-chase experiments performed with synchronized cells.[28-30] It was observed that a pulse label incorporated into DNA in the middle of the S-phase was first concentrated in the nuclear matrix DNA and then, after a chase, moved to the cleaved-off (loop DNA) fraction. This was exactly the result that one would expect knowing that the replication forks are attached to the nuclear matrix.[31,32] A different distribution of the pulse label between the matrix-bound and cleaved-off DNA fractions was, however, observed when the pulse was given at the very beginning of the S-phase (i.e., when the label was incorporated close to the replication origins). In this case the label could not be chased from the matrix-bound DNA fraction, indicating that the replication origins were located close to the matrix throughout the whole cell cycle.[28-30] Similar results were reported later by other authors.[33,34] In agreement with these results, it has been found recently that cloned nuclear matrix DNA fragments from cells in G0 and G1 phases replicate at the very beginning of the S phase indicating that they are located close to the replication origins.[35]

Independent evidence supporting the supposition that replication origins are permanently bound to the nuclear matrix was obtained in experiments on mapping the replication origin in the domain of chicken α-globin genes.[36,37] The origin was found within a 3 kb long DNA fragment of the upstream part of the domain which had been shown earlier to be permanently attached to the nuclear matrix.[36] Furthermore, the comparison in corenaturation experiments of unique DNA sequences surrounding the replication origins and permanent sites of DNA attachment to the nuclear matrix indicated that at least in chicken erythroid cells all origins were permanently attached to the nuclear matrix, i.e., were located at the DNA loop anchorage sites or very close to them.[36]

A cohabitation of replication origins and DNA loop anchorage regions was observed in several genomic areas where both replication origins and DNA loop anchorage regions were mapped. Thus, experiments on LAR mapping by topoisomerase II-mediated DNA loop excision made it possible to localize a LAR in the upstream area of the human *c-myc* gene.[38] Previously it had been demonstrated by several research groups that the replication origin was located in this area.[39-41] Cohabitation of LAR and a replication origin was also observed in a 2 kb long human genome fragment located downstream to the gene encoding lamin B2 (Lagarkova MA and Razin SV, manuscript in preparation). Recently a replication origin (initiation zone) has been mapped in proximity to the immunoglobulin heavy chain enhancer where a MAR is located.[42] In ribosomal gene clusters of different organisms, the replication origins seem to be located at or close to the DNA loop anchorage sites.[43-45] Using the multicolor fluorescence in situ hybridization technique, Bickmore and Oghene have demonstrated that in a long stretch of unique DNA on human chromosome 11 the most early replicating region is located in close proximity to the axis of the metaphase chromosome.[45]

As has been discussed above, the organization of replication origins in higher eukaryotes is not yet properly understood (section 9.1). In yeast cells, all presently mapped genomic replication origins contain ARS sequences. From the point of view of the present discussion a remarkable observation is that MARs constitute a part of a full-sized ARS[46,47] and that deletions of MARs drastically suppress the ability of ARSs to support autonomous replication of plasmids.[48] It is also worth mentioning that MARs and known replication origins of both lower and higher eukaryotes share common DNA sequence motifs including AT and TAT repeats[49] and inverted repeats.[50-56] Recently, it has been shown that short chicken DNA fragments isolated from permanent sites of DNA attachment to the nuclear matrix enhance autonomous replication of plasmids in chicken cells, although it is not yet clear whether they serve as replication origins.[57] It seems more likely that autonomous replication of plasmids bearing nuclear matrix DNA fragments is promoted, because the above fragments facilitate the incorporation of these plasmids into the preexisting organization of replication sites at the nuclear matrix. Indeed, a direct correlation between the replication efficiency of transfected circular DNA and the degree of association of this DNA with the nuclear matrix has been revealed.[57]

Taken together, all the above discussed data suggest that the attachment of replication origins to the nuclear matrix/chromosomal scaffold is one of the basic principles of the eukaryotic chromosome organization. In chapter 11 we shall discuss experimental results indicating that exogenous (viral or episomal) DNA, which replicates in eukaryotic cell nuclei, also behaves like genomic DNA loops interacting with the nuclear matrix through replication origins.

Studies of DNA-containing viruses have demonstrated that replication origins and transcriptional units frequently share same enhancer elements.[15,58,59] However, in contrast to transcriptional enhancers, which may operate at a considerable distance from the target genes, replication enhancers are able to fulfill their function only when they are located close to other functional domains of a replication origin.[60] In higher eukaryotes, replication origins are frequently located within the upstream regulatory regions of genes.[39-42,61-63] As discussed before (chapter 6), the replication origin of the chicken α-globin gene domain cohabits with the loop anchorage site and the transcription enhancer.[36,64] Similar cohabitation of nuclear matrix attachment, origin of replication and transcription enhancer activities was observed in the 5' region of the human choline acetyltransferase gene.[65] However, this is not a common rule. For example, the "classical" replication origin (*ori* β) of the Chinese hamster DHFR amplicone is located 17 kb downstream to the DHFR gene.[14] In prokaryotic systems it was directly demonstrated that activation of a replication origin located at a distance from a replication enhancer is achieved by creation of a proper three dimensional organization of DNA involving looping of DNA fragments located between the enhancer and the replication origin.[66] In cells of higher eukaryotes, association with the nuclear matrix of both replication origins and regulatory domains of different genes (see chapter 7) apparently keep all these elements close to each other and, hence, creates a three-dimensional organization of regulatory sequences permitting direct physical interaction between protein factors bound to all these sequences.

Mechanisms of the fixation of replication origins at the nuclear matrix are not yet studied. As in all other cases, it is likely that this fixation is maintained via interaction of DNA with some nuclear matrix proteins. In this respect it is worth mentioning that a protein recognizing yeast ARS consensus sequence was isolated from the extract of yeast nuclear matrix proteins.[67] One of the factors recognizing DNA sequence motives present within the replication origin of the chicken α-globin gene domain also was found to be a nuclear matrix protein.[68]

As has been discussed previously (chapter 7), the replication origins are nonrandomly distributed in the nuclear space. Recent evidence indicates that they are organized into clusters (replication foci) including 20 or more replicons.[69-72] The integrity of replication foci does not depend on the ongoing process of replication. It has been reported that replication foci may be visualized using immunostaining before the onset of replication.[73,74] Other experimental evidence indicates that replication foci persist throughout the whole cell cycle (including mitosis) in both animal and plant cells.[75,76] Keeping in mind the fact that basic DNA loops (i.e., loops which are not partitioned into sub domains as a result of transient interactions of active genes with the nuclear matrix) are bound to the matrix through replication origins, one can conclude that the loops themselves are arranged into clusters around the replication factories. These clusters may well correspond to minibands predicted by the metaphase chromosome model proposed by Coffey and collaborators.[77] According to this model, the metaphase chromosome is assembled by stacking minibands. The model is based on a simple assumption that metaphase chromosome has a regular structure. Theoretical considerations predict that in this case each miniband should contain 18 DNA loops,[77] a figure similar to the estimated number of replicons in a focus (see above). Naturally, the real situation may be more complicated. In particular, the number of replicons assembled in each cluster/factory is not likely to be the same. The sizes of individual replicons also may vary considerably. However, the assumption that a metaphase chromosome is assembled by stacking replication clusters may easily be reconciled with the above mentioned deviations from the regularity. As a matter of fact, it is known that the thickness of a metaphase chromosome is not the same throughout its whole length. Different patterns of organization and/or composition of the clusters of early-replicating and late-replicating replicons may be related to characteristic properties of R and G/Q bands in metaphase chromosomes.

A chromomeric model for the nuclear and chromosomal structure has been proposed recently by P. Cook.[78] This model also postulates that DNA loops are arranged into clusters. However the clusters are thought to be held by transcription complexes.[78] An obvious prediction from the model is that the loop length should differ significantly in euchromatic and heterochromatic regions. Some recent data clearly demonstrate that this is not the case.[45] On the contrary, the same data do not exclude the possibility that the chromosome is built from replication-related clasterosomes, as proposed by Berezney and collaborators[79,80] and by Coffey and collaborators.[77] Indeed, there are no reasons to think that the size of replicons should depend on the transcriptional status of a particular genomic area or on the cell lineage. As a matter of fact, it has been demonstrated that in different cell lines the domain of human β-globin genes replicates from the same origin independently of both the transcriptional status and the replication timing of this genomic region.[81]

9.3. CONCLUSIONS

The organization of chromosomal DNA into looped domains is directly related to the organization of replication units due to the location of replication origins at or close to the DNA loop anchorage sites. Correspondingly, the loops are likely to be organized into clusters or mini bands due to the attachment of several origins to a single replication factory.

Related Review Articles

Razin SV. DNA interactions with the nuclear matrix and spatial organization of replication and transcription. Bioessays 1987; 6:9-23.

Van der Velden HMV, Wanka F. The nuclear matrix—its role in the spatial organization and replication of eukaryotic DNA. Mol Biol Rep; 1987; 12:69-77.

Dijkwel PA, Hamlin J. Origins of replication and the nuclear matrix: The DHFR domain as a paradigm. Int Rev Cytol 1995; 162A: 455-484.

References

1. Huberman JA, Riggs AD. Autoradiography of chromosomal DNA fibers from Chinese hamster cells. Proc Natl Acad Sci USA 1966; 55:599-606.
2. Umek RM, Linskens, MHK, Kowalski D et al. New beginnings in the studies of eukaryotic DNA replication origins. Biochim Biophys Acta 1989; 1007:1-14.
3. Fangman WL, Brewer BJ. Activation of replication origins within yeast chromosomes. Annu Rev Cell Biol 1991; 7:375-402.
4. Marahrens Y, Stillman B. A yeast chromosomal origin of DNA replication defined by multiple functional elements. Science 1992; 255:817-823.
5. DePamphilis ML. Eukaryotic DNA replication: anatomy of an origin. Annu Rev Biochem 1993; 62:29-63.
6. Stinchcomb DT, Struhl K, Davis RV. Isolation and characterization of a yeast chromosomal replicator. Nature 1979; 282:39-43.
7. Biamonti G, Della Vale G, Talariko D et al. Fate of exogenous recombinant plasmids introduced into mouse and human cells. Nucl Acids Res 1985; 16:5545-5561
8. Holst A, Mueller F, Zastrow G et al. Murine genomic DNA sequences replicating autonomously in mouse L cells. Cell 1988; 52:355-365.
9. Grummt F. Autonomous replication in mouse cells: A correction. Cell 1989; 56:143-144.

10. Krysan PJ, Calos MP Replication initiates at multiple locations on an autonomously replicating plasmid in human cells. Mol Cell Biol. 1991; 11:1464-1472.
11. Muller WJ, Mueller CR, Mes AM et al. The isolation of transformed cell lines capable of supporting the replication of polyomavirus-plasmid recombinants. Mol Cell Biol 1984; 4: 2406-2412.
12. Johnson EM, Jelinek WR. Replication of plasmids bearing a human Alu-family repeat in monkey COS-7 cells. Proc Natl Acad Sci USA 1986; 83:4660-4664.
13. Hamlin JL. Mammalian origins of replication. BioEssays 1992; 14:651-659.
14. Vassilev LT, DePamhilis ML. Guide to identification of origins of DNA replication in eukaryotic cell chromosomes. Crit Rev Biochem Mol Biol 1992; 27:445-472.
15. DePamphilis ML. Origins of DNA replication in metazoan chromosomes. J Biol Chem 1993; 268:1-4.
16. Burhans WC, Vassilev LT, Caddle MS et al. Identification of an origin of bidirectional replication in mammalian chromosomes. Cell 1990; 62:955-965.
17. Vaughn JP, Dijkwel PA, Hamlin JL. Replication initiates in a broad zone in the amplified CHO dihydrofolate reductase domain. Cell 1990; 61:1075-1087.
18. Braunstein JD, Schulze D, Delguidice T et al. The temporal order of replication of immunoglobulin heavy chain constant region sequences corresponds to their linear order in the genome. Nucl Acids Res 1982; 10:6887-6902.
19. Calza RE, Eckhardt LA, Delguidice T et al. Changes in gene position are accompanied by a change in time of replication. Cell 1984; 36:689-696.
20. Harland RM, Laskey RA. Regulated replication of DNA microinjected into eggs of *X. laevis*. Cell 1980; 21:761-771.
21. Laskey RA, Kearsey SE, Mechali et al. Chromosome replication in early *Xenopus* embryos. Cold Spring Harb Symp Quant Biol 1985; 50:657-663.
22. Mechali M, Kearsey S. Lack of specific sequence requirement for DNA replication in *Xenopus* eggs compared with high sequence specificity in yeast. Cell 1984; 38:55-64.
23. Gilbert DM, Miyazavwa H, DePamphilis ML. Site-specific initiation of DNA replication in *Xenopus* egg extract requires nuclear structure. Mol Cell Biol 1995; 15:2942-2954.
24. Blow JJ, Laskey RA. Initiation of DNA replication in nuclei and purified DNA by a cell-free extract of *Xenopus* eggs. Cell 1986; 47:577-587.
25. Blow JJ, Sleeman AM. Replication of purified DNA in *Xenopus* egg extract is dependent on nuclear assembly. J Cell Sci 1990; 98:383-391.

26. Buongiorno-Nardelli M, Gioacchino M, Carri MT et al. A relationship between replicon size and supercoiled loop domains in the eukaryotic genome. Nature 1982. 298, 100-102.
27. Hartwig M. The size of folded loops in mammalian nuclear DNA. Stud Biophys 1982; 89:39-43.
28. Aelen JM, Opstelsten RJ, Wanka F. Organization of DNA replication in Physarum polycephalum. Attachment of origins of replication and replication forks to the nuclear matrix. Nucl Acids Res 1983; 11:1181-1195.
29. Van der Velden HM, Poot M, Wanka F. In vitro DNA replication in association with the nuclear matrix of permeable mammalian cells. Biochim Biophys Acta 1984; 782:429-436.
30. Van der Velden HMV, Wanka F. The nuclear matrix—its role in the spatial organization and replication of eukaryotic DNA. Mol Biol Rep 1987; 12:69-77.
31. Berezney R, Coffey DS. Nuclear protein matrix: association with newly synthesized DNA. Science 1975; 189:291-292.
32. Pardoll DM, Vogelstein B, Coffey DS. A fixed site of DNA replication in eukaryotic cells. Cell 1980; 19:527-536.
33. Carri MT, Micheli G, Graziano E et al. The relationship between chromosomal origins of replication and the nuclear matrix during the cell cycle. Exp Cell Res 1986; 164:426-436.
34. Dijkwel PA, Wenink PW, Poddighe J. Permanent attachment of replication origins to the nuclear matrix in BHK cells. Nucl Acids Res 1986; 14:3241-3249.
35. Brylawski BP, Tsongalis GJ, Cordeirostone M et al. Association of putative origins of replication with the nuclear matrix in normal human fibroblasts. Cancer Res 1993; 53:3865-3868.
36. Razin SV, Kekelidze MG, Lukanidin EM et al. Replication origins are attached to the nuclear skeleton. Nucl Acids Res 1986; 14:8189-8207.
37. Verbovaia L, Razin SV. Analysis of the replication direction through the domain of α-globin-encoding genes. Gene 1995; 166:255-259.
38. Gromova II, Thomsen B, Razin SV. Different topoisomerase II antitumour drugs direct similar specific long-range fragmentation of an amplified *c-myc* gene locus in living cells and in high-salt-extracted nuclei. Proc Natl Acad Sci USA 1995; 92:102-106.
39. Iguchi-Ariga SMM, Okazaki T, Itani T. An initiation site of DNA replication with transcriptional enhancer activity present upstream of the *c-myc* gene. EMBO J 1988; 7:3135-3142.
40. Leffak M, James CD. Opposite replication polarity of the germ line *c-myc* gene in HeLa cells compared with that of two Burkitt lymphoma cell lines. Mol Cell Biol 1989; 9:586-593.
41. Vassilev L, Johnson EM. An initiation zone of chromosomal DNA replication located upstream of the *c-myc* gene in proliferating HeLa cells. Mol Cell Biol 1990; 10:4899-4904.

42. Ariizumi K, Wang Z, Tucker PW. Immunoglobulin heavy chain enhancer is located near or in an initiation zone of chromosomal DNA replication. Proc Natl Acad Sci USA 1993; 90:3695-3699.
43. Marilley M, Gassend-Bonnet G. Supercoiled loop organization of genomic DNA: a close relationship between loop domains, expression units, and replicon organization in rDNA from *Xenopus laevis*. Exp Cell Res 1989; 180:475-489.
44. Du C, Sanzgiri RP, Shaiu WL et al. Modular structural elements in the replication origin region of *Tetrahymena* rDNA. Nucl Acids Res 1995; 23:1766-1774.
45. Bickmore WA, Oghene K. Visualizing the spatial relationships between defined DNA sequences and the axial region of extracted metaphase chromosomes. Cell 1996; 84:95-104.
46. Amati B, Gasser SM. Chromosomal ARS and CEN elements bind specifically to the yeast nuclear scaffold. Cell 1988; 54:967-978.
47. Sykes RC, Lin D, Hwang SJ et al. Yeast ARS function and nuclear matrix association coincide in a short sequence from the human *HPRT* locus. Mol Gen Genet 1988; 212:301-309.
48. Amati B, Gasser SM. *Drosophila* scaffold-attached regions bind nuclear scaffolds and can function as ARS elements in both budding and fission yeasts. Mol Cell Biol 1990; 10:5442-5454.
49. Boulikas T. Homeotic protein binding sites, origins of replication and nuclear matrix anchorage sites share the ATTA and ATTTA motifs. J Cell Biochem 1992; 50:1-13.
50. Boulikas T, Kong CF. Multitude of inverted repeats characterizes a class of anchorage sites of chromatin loops to the nuclear matrix. J Cell Biochem 1993; 53:1-12.
51. Frappier L, Zannis-Hadjopoulos M. Autonomous replication of plasmids bearing monkey DNA origin-enriched sequences. Proc Natl Acad Sci USA 1987; 84:6668-6672.
52. Zannis-Hadjopoulos M, Kaufmann G, Martin RG. Mammalian DNA enriched for replication origins is enriched for snap-back sequences. J Mol Biol 1984; 179:577-586.
53. Zannis-Hadjopoulos M, Kaufmann G, Wang SS et al. Properties of some monkey DNA sequences obtained by a procedure that enriches for DNA replication origins. Mol Cell Biol 1985; 5:1621-1629.
54. Rao BS, Zannis-Hadjopoulos M, Price GB et al. Sequence similarities among monkey ori-enriched (ors) fragments. Gene 1990; 87:233-242.
55. Bell D, Sabloff M, Zannis-Hadjopoulos M et al. Anti-cruciform DNA affinity purification of active mammalian origins of replication. Biochim Biophys Acta 1991; 1089:299-308.
56. Buolikas T. Common structural features of replication origins in all life forms. J Cell Biochem 1996; 60:297-316.

57. Bogdanova AN, Razin SV, Vassetzky YS. Nuclear matrix-associated fragments enhance autonomous replication of plasmids in chicken cells. Biochimie 1995; 77:880-887.
58. Hertz GZ, Mertz JE. Bi-directional promoter elements of SV40 are required for efficient replication of the viral DNA. Mol Cell Biol 1986; 6:3513-3522.
59. DePamphilis ML. Transcriptional elements as components of eukaryotic origins of DNA replication. Cell 1988; 52:635-638.
60. Coenjaerts FEJ, De Vries E, Pruijn GJM et al. Enhancement of DNA replication by transcription factors NFI and NFIII/Oct-1 depends critically on the positions of their binding sites in the adenovirus origin of replication. Biochim Biophys Acta 1991; 1090:61-69.
61. Taira T, Iguchi-Ariga SMM, Ariga H. A novel DNA replication origin identified in the human heat shock protein 70 gene promoter. Mol Cell Biol 1994; 14:6386-6397.
62. Zhao Y, Tsutsumi R, Yamaki M et al. Initiation zone of DNA replication at the aldolase B locus encompasses transcription promoter region. Nucl Acids Res 1994; 22:5385-5390.
63. Giacca M, Zentilin L, Norio P et al. Fine mapping of a replication origin of human DNA. Proc Natl Acad Sci USA 1994; 91: 7119-7123.
64. Razin SV, Vassetzky Jr YS, Kvartskhava AI et al. Transcriptional enhancer in the vicinity of a replication origin within the 5'-region of the chicken α-globin gene domain. J Mol Biol 1991; 217: 595-598.
65. Boulikas T, Xie L, Kong CF et al. A human regulatory sequence with nuclear matrix attachment, origin of replication, and transcription enhancer activities at the 5' region of the choline acetyltransferase gene. Submitted 1996.
66. Miron A, Mukherjee S, Bastia D. Activation of a distant replication origins in vivo by DNA looping as revealed by a novel mutant form of an initiator protein defective in cooperativity at a distance. EMBO J 1992; 11:1205-1216.
67. Hofmann JFX, Gasser SM. Identification and purification of a protein that binds the yeast ARS consensus sequence. Cell 1991; 64:951-960.
68. Vassetzky YS, De Moura Gallo CV, Bogdanova AN et al. The sequence-specific nuclear matrix-binding factor F6 is a chicken GATA-like protein. Mol Gen Genet 1993; 238:309-314.
69. Nakayasu H, Berezney R. Mapping replication sites in eukaryotic cell nucleus. J Cell Biol 1989; 108:1-11.
70. Neri LM, Mazzotti G, Capitani S et al. Nuclear matrix-bound replication sites detected in situ by 5'-bromodeoxyuridine. Histochemistry 1992; 98:19-32.

71. Hassan AB, Cook PR. Visualization of replication sites in unfixed human cells. J Cell Sci 1993; 105:541-550.
72. Hozak P, Hassan AB, Jackson DA et al. Visualization of replication factories attached to the nucleoskeleton. Cell 1993; 73:361-373.
73. Kill IR, Bridger JM, Campbell KH et al. The timing of the formation and usage of replitase clusters in S-phase nuclei of human diploid fibroblasts. J Cell Sci 1991; 100:869-876.
74. Adachi Y, Laemmli UK. Identification of nuclear pre-replication centers poised for DNA synthesis in *Xenopus* egg extracts: Immunolocalization study of replication protein A. J Cell Biol 1992; 119:1-15.
75. Meng C, Berezney R. Replication cluster domains persist throughout the cell cycle of mouse 3T3 cells. J Cell Biol 1991; 115:95a
76. Sparvoli E, Levi M, Rossi E. Replicon clusters may form structurally stable complexes of chromatin and chromosomes. J Cell Sci 1994; 107:3097-3103.
77. Pienta KJ, Getzenberg RH, Coffey DS. Cell structure and DNA organization. Crit Rev Eukar Gene Expres 1991; 1:355-385.
78. Cook PR. A chromomeric model for nuclear and chromosome structure. J Cell Sci 1995; 108:2927-2935.
79. Tubo RA, Berezney R. Identification of 100 and 150 S DNA polymerase alpha-primase megacomplexes solubilized from the nuclear matrix of regenerating rat liver. J Biol Chem 1987; 262:5857-5865.
80. Berezney R, Mortillaro MJ, Ma H et al. The nuclear matrix: a structural milieu for genomic function. Int Rev Cytol 1995; 162A:1-65.
81. Kitsberg D, Selig S, Keshet I et al. Replication structure of the human β-globin gene domain. Nature 1993; 366:588-590.

CHAPTER 10

Do DNA Loop Anchorage Sites Define the Borders of Transcription Units or Transcriptionally Active Genomic Regions?

10.1. TOPOGRAPHICAL APPROACH

There are two approaches to answer the question posed in the title of this chapter. The first approach can be referred to as a topographical one. Indeed, it is possible simply to map the loop anchorage sites in different genomic areas and to see whether they frame transcription units. The results will of course depend on the method used for mapping the anchorage sites. In spite of the fact that MARs and SARs constitute the sites of in vitro DNA binding to the nuclear matrix, they are still frequently considered as the DNA loop borders (see discussion in chapter 5). Some MARs/SARs have indeed been found at the borders of DNase I-sensitive (i.e., transcriptionally active) genomic domains.[1-3] However, they can be found also within genes.[4-6] Hence, the statement that MARs/SARs delimit the borders of transcriptionally active genomic domains would not be correct in general. Recent experiments on mapping loop anchorage sites in a 500 kb region of the *Drosophila* first chromosome have shown, however, that only a fraction of MARs/SARs may actually be involved in the anchorage of

DNA loops on the nuclear matrix.[7] As far as LARs mapped by topoisomerase II-mediated DNA loop excision are concerned, there is a clear shortage of mapping data to make a final conclusion. Yet, at least in one case a LAR was found within a gene.[7] It also seems obvious that some genes of higher eukaryotes are too long to be organized in one loop (for example, the human Dystrophin gene is more than 2,000 kb long). Our preliminary data suggest that these genes are organized into several loops (Razin SV et al, unpublished). Hence it is likely that transcriptional complexes can pass somehow the loop anchorage sites. This possibility may stem from a complex organization of LARs involving multiple interactions of DNA with the nuclear matrix proteins. It was proposed previously by Garrard and collaborators[8] that the presence of multiple binding sites within a matrix attachment region may allow the passage of DNA and RNA polymerases without disrupting the loop organization.

It will be interesting perhaps to consider the evolutionary aspect of the loop ends distribution in the genome. It is obvious that the evolutionary age of various genes and gene clusters in the mammalian genome differs significantly. It is possible that the most ancient genes (for example, genes coding for ribosomal RNAs) are located within single loops. Long multi exon genes apparently represent a relatively modern product of the evolution. These genes have probably originated as a result of combining and repositioning of genomic domains (see discussion in chapter 11). The organization of these genes into several loops may hence reflect the way of their generation. In this regard it is noteworthy that intergenic MARs were found only in introns.[9]

10.2. FUNCTIONAL APPROACH

The second approach that can be used to find out whether DNA loop anchorage sites define the borders of active genomic domains may be referred to as a functional one. It is based on the construction of artificial minidomains insulating the reporter genes from position variegation effects in the genome of transgenic animals. It is clear that in differentiated mammalian cells most of genomic domains are not active. Those which are active can be recognized experimentally due to their preferential sensitivity to nucleases.[10] Whatever the nature of chromatin domains identified by preferential sensitivity to nucleases may be, there are good reasons to believe that these domains constitute relatively independent functional units of the genome. The transcriptional status of a domain as a whole seems to be under control of special regulatory mechanisms (see below and also refs. 11, 12). When exogenous DNA is microinjected into the pronuclei of zygotes, its integration into any genomic domain seems to be equally probable. In differentiated cells of an adult organism, evidently only genes that happen to integrate into domains active in a given cell type will be expressed (Fig. 10.1). A logical way to solve the problem of position effects is to create artificial minidomains including a foreign gene and a block of regulatory sequences. The active domains in chromatin characterized by preferential sensitivity to DNase I usually have relatively sharp borders. This suggests that there should be some special genomic elements defining these borders. So, the questions arise: what are these elements and can they be found among DNA sequences located at the basements of DNA loops? A straight experimental approach to answer the above questions consists in the creation of artificial genomic domains containing a reporter gene surrounded by putative domain border elements. In a positive case, the above elements are expected to protect the reporter gene from position variegation effects (Fig. 10.1). In other words, the expression of this gene in different tissues of transgenic animals should be position-independent and copy number-dependent. Up to now, only MARs/SARs have been tested for the ability to insulate transgenes from suppressing effects of the host chromatin domain. In several experiments made by different authors it was observed that MARs, when placed upstream and downstream to the transgene, suppressed (in some cases only partially) the position variegation effects.[13-18] For presently unknown reasons, the ability of MARs to suppress position effects seems to be especially pronounced in plant cells.[15,18-21] The level of activation of transgene expression within minidomains defined by MARs/SARs varied considerably from 10-fold[13-14] to 200-fold.[17] In contrast to the stimulation of gene expression by enhancers, which may be observed in transient assay (i.e., when transcription of episomal genes is studied), MARs stimulated gene expression only after stable integration of corresponding constructs into the host-cell genome.

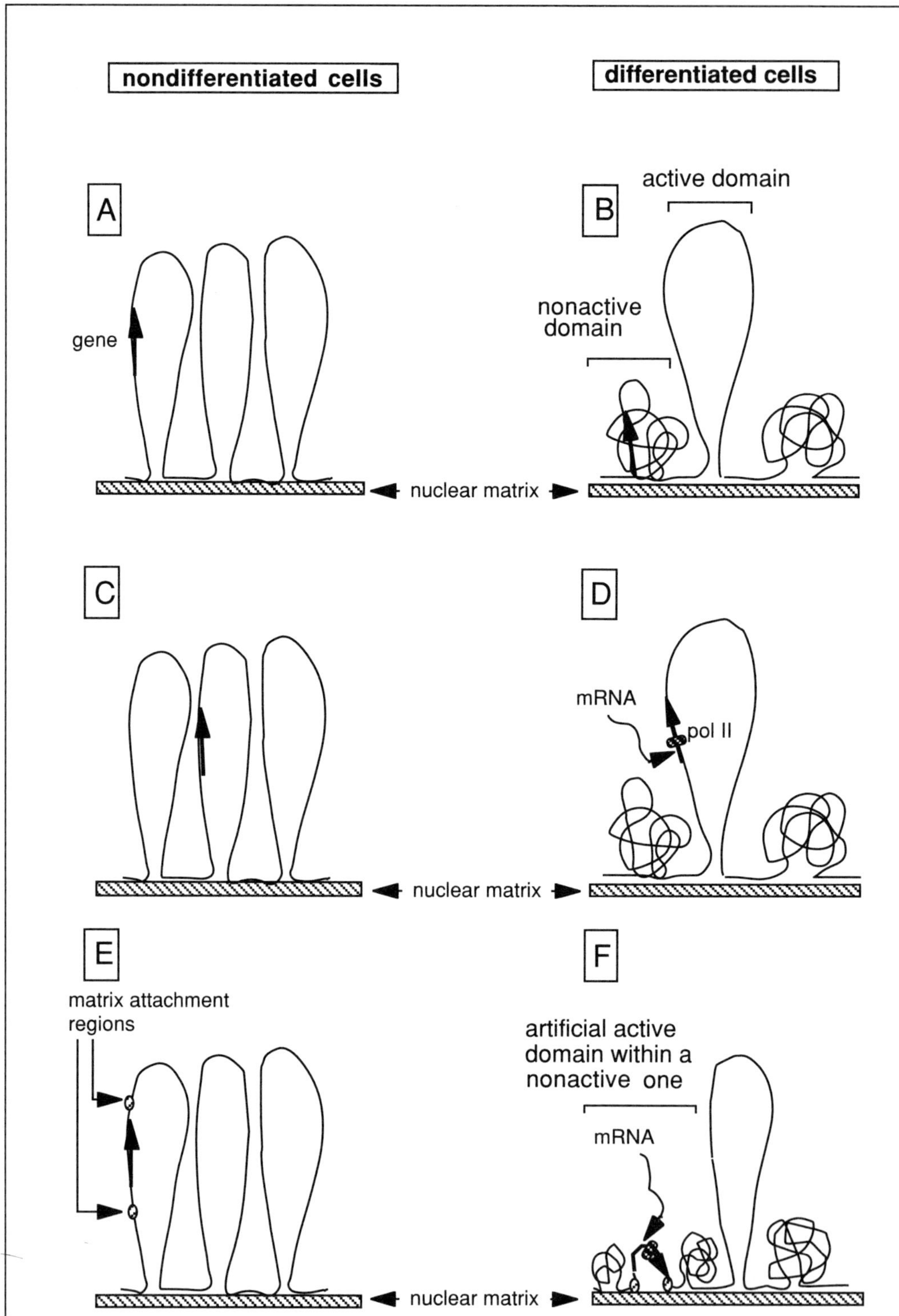
nondifferentiated cells
differentiated cells
A
gene
B
active domain
nonactive domain
nuclear matrix
C
D
mRNA
pol II
nuclear matrix
E
matrix attachment regions
F
artificial active domain within a nonactive one
mRNA
nuclear matrix

It should be stressed that in experiments with transgenic animals and cell lines bearing foreign genes integrated into the genome it is not easy to distinguish protection from the position effects and simple stimulation of a transgene expression by MARs. A careful analysis carried out in the laboratory of Laemmli has justified a conclusion that MARs flanking a reporter gene stimulate the expression of this gene 20- to 40-fold but do not dampen the position variegation effects.[22] Similar observations were made by other authors.[23-27] For example, it was demonstrated that MARs/SARs ensured copy-dependent expression of a reporter gene in nondifferentiated cells of a mouse embryo, but not in differentiated cells of newborn or adult mice.[24] It was proposed that the stimulatory effect of MARs/SARs on transcription was due to the ability of these elements to serve as nucleation sites for the histone H1.[28,29] According to this proposition, MARs/SARs can stimulate or suppress transcription depending on the ratio of H1 to displacement proteins (such as HMG-I/Y) capable of removing H1 from MARs/SARs. Correspondingly, MARs/SARs are regarded as elements involved in controlling the status (open or closed) of chromatin domains. However, the proposed model[28,29] does not explain why the stimulatory effect of MARs/SARs cannot be observed in transient assay.

Some indirect evidence suggests that MARs can create a closed domain capable of maintaining nonconstrained superhelical tension. The ability of MAR elements to fulfill the function of a topological lock may be explained by their high unwinding potential and capacity to be organized into cruciforms and other noncanonical secondary structures (see section 6.2 and also ref. 30). Unpaired DNA regions may be fixed via their interaction with some nuclear matrix proteins. Obviously, this will prevent the

Fig. 10.1 (opposite). A scheme illustrating the strategy of dampening position effects by creation of minidomains in the genome of transgenic animals. In embryonic (nondifferentiated) cells (A, C, E) all domains are open. The probability of integration of a transgene into each domain is supposed to be similar or even equal. In differentiated cells the majority of genomic domains becomes nonactive (condensed loops in sections B, D, F). In a nonactive domain a transgene is not expressed (B). The transgene happened to reside in an active domain is expressed (D). In order to insulate a transgene from the influence of the host-cell chromatin the transgene is places between matrix attachment regions (E). It is supposed that interaction of the latter with the nuclear matrix should create an independent domain (F). Within the above domain expression of the transgene will be determined only by regulatory elements placed into this domain along with the transgene. In a result expression of the transgene will not depend on the site of integration.

possibility of their rotation. In a closed chromatin domain, negative supercoils may accumulate as a result of acetylation of histones which cause changes in the parameters of nucleosomes triggering a partial release of DNA supercoils constrained by the nucleosomes.[31,32] Negative supercoiling of DNA is thought to facilitate transcription (reviewed by Freeman and Garrard[33]). The histones of transcriptionally active chromatin are known to be hyperacetylated.[34-37] Hyperacetylation of histones H3 and H4 can be induced experimentally by treating cells with butyrate[38] or other inhibitors of histone deacetylases. It was reported that this treatment markedly increased the level of transcription of a reporter gene flanked by MAR sequences.[39,40] The effect of the same treatment on the expression of a marker gene nonflanked by MARs was very weak (if at all detectable). The simplest interpretation of the above results is that MARs create an independent small closed domain within which the superhelical tension released as a result of acetylation of nucleosomal histones can be accumulated.[39,40] Some data indicate that nonconstrained DNA supercoiling is essential for the establishment and maintenance of DNase sensitive (i.e., transcriptionally active) chromatin domains.[41-43] Hence, it is easy to explain why elements restricting the spreading of superhelical tension (MARs/SARs) are located at the borders of these domains.[1-3] Partitioning (perhaps temporal) of large closed domains into smaller ones mediated by internal MARs/SARs may also be of some biological importance. It was proposed that local positive supercoiling generating ahead of a transcription complex (see chapter 7) facilitates transcription through nucleosomes (for a review see refs. 44-46). It is possible that superhelical tension can be retained for a longer time in short closed domains. If domains of this kind may be created by MARs/SARs,[39,40] the stimulation of transcription by MARs/SARs[22-26] can be easily explained. The MARs located within the transcription units were also proposed to serve as molecular switches capable of changing their structure (for example, melt or form cruciforms, triplexes etc.) under certain degree of superhelical tension.[47]

One problem remained unrecognized in the initial experiments with minidomains, arises from the multiple integration of transgenes in the same genomic position. Recent observations suggest that tandem arrays of multicopy transfectants may become transcriptionally inactive in spite of the presence of all necessary regulatory elements.[17,48,49] Hence, the data obtained in experiments with a single copy integration or a multiple integration in different positions are more reliable (see ref. 47 for a discussion). A certain degree of discrepancy in the results of experiments on creation artificial minidomains may be due to the lack of understanding of mechanisms regulating expression of genes at the domain level. It has been mentioned that to protect a transgene from position effects a MAR should operate in combination with an enhancer.[16,47] In many experiments with minidomains defined by MARs, marker genes were placed under control of heterologous promoters and enhancers (see ref. 47 and references therein). Clearly, this is an oversimplified model of natural genomic domains. The latter may contain a number of authentic regulatory sequences which cooperate with boundary elements in establishing active chromatin configuration and dampening position effects upon integration of the whole domain into a genome of heterologous cells.[50-53]

In spite of some contradictions, the above discussed results suggest that MARs/SARs (which are likely to constitute an essential part of DNA loop anchorage regions) can function as the boundaries of functional chromatin domains. Correspondingly, these domains can be identified with DNA loops fixed at the nuclear matrix. This tentative conclusion was, however, questioned when another approach was used for identification of the borders of functional chromatin domains. This approach known as "enhancer-blocking assay" was developed by Schedl and collaborators.[54,55] It was assumed that upon integration into genomic DNA the domain border should insulate the reporter gene not only from the negative influence of the host cell chromatin but also from the positive influence of an enhancer (Fig. 10.2). Being placed between the reporter gene and the enhancer such element may be expected to block the action of the enhancer (upon stable integration of the whole construct in the host cell genomic DNA). With this approach it was found that MAR/SAR from the intergenic spacer separating two *hsp 70* genes in *Drosophila* 87A7 locus did not behave as an insulator[55] while some genomic elements containing no MARs/SARs showed the properties of insulators in the enhancer-blocking assay.[55] On the other hand, the MAR from DNase I HS 5 of the human β-globin domain locus control region[56]

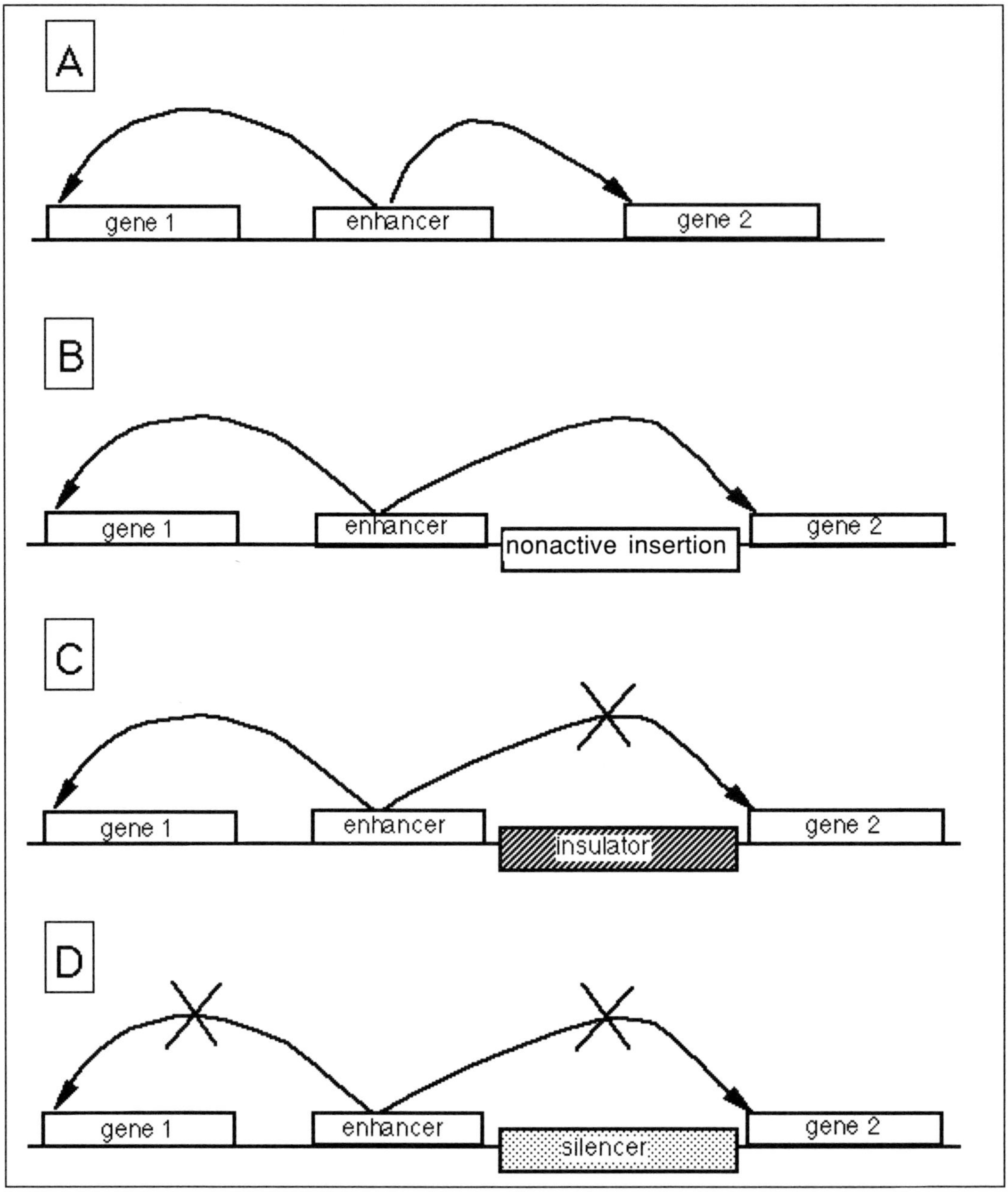

Fig. 10.2. A scheme illustrating the strategy of identification insulators by enhancer-blocking assay. (A) A construct containing an enhancer placed between two reporter genes is shown. After integration of this construct into the cellular genome the enhancer will stimulate expression of both genes. (B) Insertion of a reasonably long indifferent DNA sequence between the enhancer an the gene 2 may be expected to have no negative effect on the expression of this gene. (C) Being inserted between the enhancer and the gene 2 an insulator functionally separates this gene from the enhancer. In a result expression of this gene will be suppressed or even stopped while the rate of the expression of gene 1 will not be changed. In contrast, insertion of silencer in the same position (section "D") should suppress the expression of both reporter genes.

seems to be able to insulate reporter genes from both positive and negative regulatory elements.[16,57] The apparent contradiction in the data concerning possible functions of MARs/SARs as the borders of functional chromatin domains can have two explanations.

First, all MARs/SARs able to confer position-independent expression to transgenes were taken from the actual borders of DNase I sensitive domains.[1-3] It is possible that these MARs/SARs possess some additional properties which make them different from MARs/SARs found within DNase I sensitive domains or even within the genes. It is also possible that relatively long DNA fragments from the domain borders may include some other essential elements of the loop anchorage sites (in addition to the MARs/SARs being mapped for these fragments). Whatever the reason of functional heterogeneity of MAR-containing DNA fragments may be, it is clear that intronic MARs influence only slightly the level of transgene expression after stable integration into the host cell genome.[58,59] The MAR/SAR possessing no insulating properties in the enhancer-blocking assay[55] was taken from the middle of the DNase I sensitive domain. The MAR/SAR from the border of the human β-globin gene domain worked as an insulator in the enhancer blocking assay.[57] It is also worth mentioning that "classical" insulators flanking the domain of *Drosophila hsp70* genes do not work in mammalian cells while the insulator isolated from the upstream border of the chicken β-globin gene domain possessed a normal functional activity in avian, mammalian and *Drosophila* cells.[60] Hence, there may be several different types of insulators.

The second possibility which we would like to discuss here is that the domain borders may not necessarily insulate a gene from positive regulatory elements. Indeed, the long term repression of different genomic domains achieved by the condensation of chromatin within DNA loops may be under the control of some negative signals favoring a more compact organization of a major part of genomic DNA. Repressive heterochromatin domains can spread over long distances entering and modifying the adjacent euchromatic regions, which is the well-known cause of the position variegation effect.[61-63] Once heterochromatin spreading is blocked by a LAR, positive regulatory elements of different types (all operating within relatively short distances) may create a decompactized chromatin structure favoring transcription (see discussion in ref.

64). There is, perhaps, no special need to restrict the action of positive regulatory elements which operate within a relatively short range.

It may be argued that Locus Control Regions (LCR)[65-70] are dominant positive regulatory elements able to influence the chromatin organization over long distances. Yet the mode of action of LCRs is not yet properly understood. The best studied LCR from the domain of human β-globin genes[65] includes several enhancers (HS 2-4) and a MAR/insulator (HS5). A change in the chromatin structure and replication timing of the whole locus was observed when the basic part of the LCR including the insulator (HS 2-5) was deleted.[71] At the same time, the insertion of the hygromycin gene within the LCR suppressed the expression of the β-globin gene but did not cause a global rearrangement in the chromatin structure of the β-globin locus.[72] Hence, it is clear that a full-sized LCR regulates the expression of β-globin genes at two levels: by opening the large chromatin domain and via enhancer-promoter interactions. Although some parts of the β-globin locus LCR (especially HS 2) were reported to suppress the position variegation effects in transgenic animals, the expression of transgenes linked to the truncated LCR was not totally position independent. The range of expression varied in different animals and the levels of human β-globin mRNA were not absolutely correlated with the transgene copy number.[73] It is possible that the MAR/insulator (HS 5) is necessary for the regulation at the level of chromatin domain. Further experiments will, perhaps, clarify the situation.

10.3. CONCLUSIONS

The relation between chromosomal DNA loops and transcriptional units or active genomic domains remains unclear. The data of mapping experiments indicate that both MARs/SARs and LARs can be found within transcription units. So, at least some DNA loop anchorage regions do not colocalize with the borders of active genomic domains. Yet, certain MARs/SARs were found to possess the functional properties of domain border sequences, i.e., the ability to insulate a transgene from suppressing effects of the host chromatin domain. As a matter of fact, the present data do not exclude the possibility that all borders of chromatin domains colocalize with the loop anchorage sites, although an opposite statement would not be correct.

Related Review Articles

Dillon N, Grosveld F. Transcriptional regulation of multigene loci: multilevel control. Trends Genet 1993; 9:134-137.

Felsenfeld G. Chromatin structure and expression of globin-encoding genes. Gene 1993; 135:119-124.

Wolffe AP. Gene regulation. Insulating chromatin. Curr Biol 1994; 4:85-87.

Bode J, Schlake T, Rios-Ramirez M, Mielke C, Stengert M, Kay V and Klehr-Wirth D. Scaffold/matrix-attached regions: Structural properties creating transcriptionally active loci. Int Rev Cytol 1995; 162A:389-454.

Bode J, Stengert-Iber M, Kay V et al. Scaffold/matrix-attached regions: Topological switches with multiple regulatory functions. Crit Rev Eukar Gene Expr 1996; 6:115-138.

Bonifer C, Huber MC, Faust N et al. Regulation of the chicken lysozyme locus in transgenic mice. Crit Rev Eukar Gene Expr 1996; 6:285-297.

References

1. Phi-Van L, Stratling WH. The matrix attachment regions of the chicken lysozyme gene comap with the boundaries of the chromatin domain. EMBO J 1988; 7:655-664.
2. Bode J, Maass K. Chromatin domain surrounding the human interferon-β gene as defined by scaffold attached regions. Biochemistry 1988; 27:4706-4711.
3. Levy-Wilson B, Fortier C. The limits of the DNase I-sensitive domain in the human apolipoprotein B gene coincide with the locations of chromosomal anchorage loops and define the 5' and 3' boundaries of the gene. J Biol Chem 1989; 264:21196-21204.
4. Ito T, Sakaki Y. Nuclear matrix association regions of rat alpha 2-macroglobulin gene. Biochem Biophys Res Commun 1987; 149:449-454.
5. Kas E, Chasin LA. Anchorage of the Chinese hamster dihydrofolate reductase gene to the nuclear scaffold occurs in an intragenic region. J Mol Biol 1987; 198:577-692.
6. Romig H, Ruff J, Fackelmayer FO et al. Characterization of two intronic nuclear-matrix-attachment regions in the human DNA topoisomerase I gene. Eur J Biochem 1994; 221:411-419.
7. Iarovaia OV, Hancock R, Lagarkova MA et al. Mapping of genomic DNA loop organization in a 500-kilobase region of the *Drosophila* X chromosome using the topoisomerase II-mediated DNA loop excision protocol. Mol Cell Biol 1996; 16:302-308.
8. Blasquez V, Sperry AO, Cockerill PN et al. Protein: DNA interactions at chromosomal loop attachment sites. Genome 1989; 31, 503-509.

9. Gasser SM, Amati BB, Cardenas ME et al. Studies on scaffold attachment sites and their relation to genome function. Int Rev Cytol 1989; 119:57-96.
10. Weintraub H, Groudine M. Chromosomal subunits in active genes have an altered conformation. Science 1976; 73:848-856.
11. Bodnar JW. A domain model for eukaryotic DNA organization: A molecular basis for cell differentiation and chromosome evolution. J Theor Biol 1988; 132:479-507.
12. Goldman MA. The chromatin domains as a unit of gene regulation. Bioessays 1988; 9:50-55.
13. Stief AC, Winter DM, Stratling WH et al. A nuclear DNA attachment element mediates elevated and position independent gene activity. Nature 1989; 341:343-345.
14. Phi-Van L, von Kries JP, Ostertag W et al. The chicken lysozyme 5' matrix attachment region increases transcription from a heterologous promoter in heterologous cells and dampens position effects on the expression of transfected genes. Mol Cell Biol 1990; 10:2302-2307.
15. Breyne P, van Montagu M, De Picker A et al. Characterization of a plant scaffold attachment region in a DNA fragment that normalizes transgene expression in tobacco. Plant Cell 1992; 4:463-471.
16. Yu J, Bock JH, Slightom JL et al. A 5' matrix-attachment region and the polyoma enhancer together confer position-independent transcription. Gene 1994; 139:139-145.
17. Kalos M, Fournier RE. Position independent transgene expression mediated by boundary elements from the apolipoprotein B chromatin domain. Mol Cell Biol 1995; 15:198-207.
18. Schoffl F, Schroder G, Kliem M et al. An SAR-sequence containing 395 bp-DNA fragment mediates enhanced, gene-dosage-correlated expression of a chimaeric heat shock gene in transgenic tobacco plants. Transgenic Res 1993; 2:93-100.
19. Mlinarova L, Loonen A, Heldens J et al. Reduced position effect in mature transgenic plants conferred by the chicken lysozyme matrix-associated region. Plant Cell 1994; 6:417-426.
20. Mlynarova L, Jansen RC, Conner AJ et al. The MAR-mediated reduction in position effect can be uncoupled from copy number-dependent expression in transgenic plants. Plant Cell 1995; 7:599-609.
21. Allen GC, Hall GE Jr, Childs LC et al. Scaffold attachment regions increase reporter gene expression in stably transformed plant cells. Plant Cell 1993; 5:603-613.
22. Poljak L, Seum C, Mattioni T et al. SARs stimulate but do not confer position independent gene expression. Nucl Acids Res 1994; 22:4386-4394.
23. Klehr D, Maass K, Bode J. Scaffold-attached regions from the human interferon B domain can be used to enhance the stable ex-

pression of genes under the control of various promoters. Biochemistry 1991; 30:1264-1270.
24. Thompson EM, Christians E, Stinnakre M-G et al. Scaffold attachment regions stimulate *hsp70.1* expression in mouse preimplantation embryos but not in differentiated tissues. Mol Cell Biol 1994; 14:4694-4703.
25. McKnight RA, Shamay A, Sankaran L et al. Matrix-attachment regions can impart position-independent regulation of a tissue-specific gene in transgenic mice. Proc Natl Acad Sci USA 1992; 89:6943-6947.
26. Brooks AR, Nagy BP, Taylor S et al. Sequences containing the second intron-enhancer are essential for transcription of the human apolipoprotein B gene in the livers of transgenic mice. Mol cell Biol 1994; 14:2243-2256.
27. Forrester WC, van Genderen C, Jenuwein T et al. Dependence of enhancer-mediated transcription of the immunoglobulin μ gene on nuclear matrix attachment regions. Science 1994; 265:1221-1225.
28. Kas E, Poljak L, Adachi Y et al. A model for chromatin opening: stimulation of topoisomerase II and restriction enzymes cleavage of chromatin by distamycin. EMBO J 1993; 12:115-126.
29. Zhao K, Kas E, Gonzalez E et al. SAR-dependent mobilization of histone H1 by HMG-I/Y in vitro: HMG-I/Y is enriched in H1-depleted chromatin. EMBO J 1993; 12:3237-3247.
30. Schubeler D, Mielke C, Maass K et al. Scaffold/Matrix-attached regions (S/MARs) act upon transcription in a context-dependent manner. Biochemistry 1996; (in press).
31. Norton VG, Marvin KW, Yau P et al. Nucleosome linking number change controlled by acetylation of histones H3 and H4. J Biol Chem 1990; 265:19848-19852.
32. Thomsen B, Bendixen C, Westergaard O. Histone hyperacetylation is accompanied by changes in DNA topology in vivo. Eur J Biochem 1991; 201:107-111.
33. Freeman LA, Garrard WT. DNA supercoiling in chromatin structure and gene expression. Crit Rev Eukar Gene Expr 1992; 2:165-209.
34. Hebbes TR, Thorne AW, Crane Robinson C. A direct link between core histone acetylation and transcroptionally active chromatin. EMBO J 1988; 7:1395-1402.
35. Boffa LC, Walker J, Chen TA et al. Factors affecting nucleosome structure in transcriptionally active chromatin. Histone acetylation, nascent RNA and inhibitors of RNA synthesis. Eur J Biochem 1990; 194:811-823.
36. Jeppesen P, Turner BM. The inactive X chromosome in female mammals is distinguished by lack of histone H4 acetylation, a cytogenetic marker for gene expression. Cell 1993; 74:281-289.
37. Hebbes TR, Clayton AL, Thorne AW et al. Core histone hyperacetylation co-maps with generalized DNase I sensitivity in

chicken b-globin chromosomal domain. EMBO J 1994; 13: 1823-1830.
38. Candido EPM, Reeves R, Davie JR. Sodium butirate inhibits histone deacetylation in cultured cells. Cell 1978; 14:105-113.
39. Klehr D, Schlake T, Maass K et al. Scaffold-attached regions (SAR elements) mediate transcriptional effects due to butirate. Biochemistry 1992; 31:3222-3229.
40. Schlake T, Klehr-Wirth D, Yoshida M et al. Gene expression within a chromatin domain: The role of core histone hyperacetylation. Biochemistry 1994; 33:4197-4206.
41. Villeponteau B, Lundell M, Martisson H. Torsional stress promotes the DNase I sensitivity of active genes. Cell 1984; 39:469-478.
42. Villeponteau B, Pribyl TM, Grant MH et al. Novobiocin induces the in vivo cleavage of active gene sequences in intact cells. J Biol Chem 1986; 261:10359-10365.
43. Villeponteau B, Martinson HG. Gamma rays and bleomycin nick DNA and reverse the DNase I sensitivity of beta-globin gene chromatin in vivo. Mol cell Biol 1987; 7:1917-1924.
44. Kornberg RD, Lorch Y. Irresistible force meets immovable object: transcription and the nucleosome. Cell 1991; 67:833-836.
45. Thoma F. Structural changes in nucleosomes during transcription: strip, split or flip? Trends Genet 1991; 7:175-177.
46. Wolffe A. Nucleosome positioning and modification: Chromatin structures that potentiate transcription. Trends Biochem Sci 1994; 19:240-244.
47. Bode J, Stengert-Iber M, Kay V et al. Scaffold/matrix-attached regions: Topological switches with multiple regulatory functions. Crit Rev Eukar Gene Expr 1996; 6:115-138.
48. Dorer DR, Henikoff S. Expansions of transgene repeats cause heterochromatin formation and gene silencing in *Drosophila*. Cell 1994; 77:993-1002.
49. Mielke C, Maass K, Tummler M, Bode J. Anatomy of highly expressed chromosomal sites targeted by retroviral vectors. Biochemistry 1996; 35:2239-2252.
50. Bonifer C, Vidal M, Grossveld F et al. Tissue specific and position independent expression of the complete gene domain for chicken lysozyme in transgenic mice. EMBO J 1990; 9:2843-2848.
51. Bonifer C, Hecht A, Saueressig H et al. Dynamic chromatin: the regulatory domain organization of eukaryotic gene loci. J Cell Biochem 1991; 47:99-108.
52. Sippel AE, Schäfer G, Faust N et al. Chromatin domains constitute regulatory units for the control of eukaryotic genes. Cold Spring Harb Symp Quant Biol 1993; 58:37-44.
53. Bonifer C, Yannoutsos N, Krüger G et al. Dissection of the locus control function located on the chicken lysozyme gene domain in transgenic mice. Nucl Acids Res 1994; 22:4202-4210.

54. Kellum R, Schedl P. A position-effect assay for boundaries of higher order chromosomal domains. Cell 1991; 64:941-950.
55. Kellum R, Schedl P. A group of scs elements function as domain boundaries in an enhancer-blocking assay. Mol Cell Biol 1992; 12:2424-2431.
56. Jarman AP, Higgs DR. Nuclear scaffold attachment sites in the human globin gene complexes. EMBO J 1988; 7:3337-3344.
57. Li Q, Stamatoyannopoulos G. Hypersensitive site 5 of the human *b* globin locus control region functions as a chromatin insulator. Blood 1994; 84:1399-1401.
58. Blasquez VC, Xu M, Moses SC et al. Immunoglobulin kappa gene expression after stable integration. I. Role of the intronic MAR and enhancer in plasmacytoma cells. J Biol Chem 1989; 264: 21183-21189.
59. Xu M, Hammer RE, Blasquez VC et al. Immunoglobulin kappa gene expression after stable integration. II. Role of the intronic MAR and enhancer in transgenic mice. J Biol Chem 1989; 264: 21190-21195.
60. Chung JH, Whiteley M, Felsenfeld G (1993) A 5' element of the chicken beta-globin domain serves as an insulator in human erythroid cells and protects against position effect in *Drosophila.* Cell 73: 505-514.
61. Reuter G, Spierer P. Position effect variegation and chromatin proteins. Bioessays 1992; 14:605-612.
62. Gatti M, Pimpinelli S. Functional elements of *Drosophila melanogaster* heterochromatin. Annu Rev Genet 1992; 26:239-275.
63. Shaffer CD, Wallrath LL, Elgin SC. Regulating genes by packaging domains: bits of heterochromatin in euchromatin? Trends Genet 1993; 9:35-37.
64. Allen GC, Hall Jr GE, Childs LC et al. Scaffold attachment regions increase reporter gene expression in stably transformed plant cells. Plant Cell 1993; 5:603-613.
65. Grosveld F, Blom van Assendelft M, Greaves DR et al. Position independent high level expression of the human β-globin gene in transgenic mice. Cell 1987; 51:975-985.
66. Lang G, Wotton D, Owen MJ et al. The structure of human CD2 gene and its expression in transgenic mice. EMBO J 1988; 7:1675-1682.
67. Felsenfeld G. Chromatin is an essential part of the transcriptional mechanism. Nature 1992; 355:219-224.
68. Carson S, Wiles MV. Far upstream regions of class II MHC Ea are necessary for position-independent, copy-dependent expression of Ea transgene. Nucl Acids Res 1993; 21:2065-2072.
69. Diaz P, Cado D, Winoto A. A locus control region in the T-cell receptor alpha/delta locus. Immunity 1994; 1:207-217.
70. Gourdon G, Sharpe JA, Wells D et al. Analysis of a 70 kb segment of DNA containing the human ζ and a-globin genes linked

to their regulatory element (HS-40) in transgenic mice. Nucl Acids Res 1994; 22:4139-4147.
71. Forrester WC, Epner E, Driscoll MC et al. A deletion of the human β-globin locus activation region causes a major alteration in chromatin structure and replication across the entire β-globin locus. Gen Dev 1990; 4:1637-1649.
72. Kim CG, Epner EM, Forrester WC et al. Inactivation of the human β-globin gene by targeted insertion into the β-globin locus control region. Gen Dev 1992; 6:928-938.
73. Ryan TM, Behringer RR, Martin NC et al. A single erythroid-specific DNase I super-hypersensitive site activates high levels of human β-globin gene expression in transgenic mice. Gen Dev 1989; 3:314-323.

CHAPTER 11

DNA Loops as Basic Units of the Genome Organization and Evolution

11.1. DNA Loop Anchorage Sites as Preferential Places of DNA Recombination Resulting in Chromosomal Rearrangements

It has long been known that chromosomal rearrangements frequently occur at some specific places ("hot spots") in the genome. As shown by analysis of the long-range distribution of "hot spots" of chromosomal DNA rearrangement, they are usually separated by 20-100 kb pieces of DNA which are rarely involved in rearrangements.[1-3] A correlation between the above distances and the average size of topological DNA loops seems to be suggestive.

Analysis of recombination junction sequences revealed frequent presence of MARs and motifs preferentially cleaved by eukaryotic topoisomerase II,[4-10] which again suggests that recombinations may occur preferentially at DNA loop anchorage sites. In agreement with the above supposition, it was found that another recombination process, namely the integration of retroviral DNA into the host-cell genome, occurred predominantly, if not exclusively, in a subset of MAR sequences.[11] Although the mechanisms of retroviral DNA integration are distinct from the mechanisms of large-scale chromosomal rearrangement, the targeting of different recombination events to the MAR sequences is worthy of notice.

Using a model system, Gale and Osheroff[12] have demonstrated that eukaryotic topoisomerase II possesses an intrinsic intermolecular DNA ligation activity and is able to mediate illegitimate DNA recombination in vitro. Exposure of mammalian cells to topoisomerase II-specific drugs was reported to stimulate different genomic rearrangements such as deletions, insertions and translocations.[13-15] Numerous observations indicate that chemotherapy of tumors with topoisomerase II-specific drugs (such as VP16 or m-AMSA) frequently causes secondary leukemias originating as a result of chromosomal rearrangements.[16-21] Again, these rearrangements occur nonrandomly in the genome. The recombination points show a high degree of clusterization. For example, the breakpoints of a number of translocations involving the MLL (myeloid-lymphoid-leukemia) gene were mapped within a 8.3 kb DNA fragment. In model systems (plasmids propagating in eukaryotic cells and simian virus 40 DNA in infected cells), the topoisomerase II-specific drugs also induced rearrangements in regions known to be preferentially cleaved by topoisomerase II.[22,23] As shown previously,[24-26] the nuclear matrix topoisomerase II (i.e., the enzyme which interacts with DNA in loop anchorage regions) constitutes a primary target for different antitumor drugs.

The above observations suggest that the loop anchorage regions may constitute preferential sites for illegitimate DNA recombination resulting either in the loss or in repositioning of large fragments of chromosomal DNA (namely of DNA loops). Although separated by long pieces of loop DNA, the loop anchorage regions are likely to be located close to each other in the nuclear space. This may further promote the possibility of recombination events between these regions. Indeed, as demonstrated in a model experiment the association of circular DNA plasmids with the nuclear matrix facilitates the topoisomerase II-mediated incorporation of these plasmids into catenated networks.[27]

Topoisomerase II is not the only enzyme which may promote recombination events between loop anchorage regions. A DNA recombinase activity with as yet poorly studied properties has been found recently in nuclear matrix preparations.[28,29] It is also worth mentioning that double stranded DNA breaks may initiate a chain of complex events resulting in deletions and translocations.[30] The loop anchorage regions represent weak spots in chromosomes as they are preferentially accessible to nucleases (see chapter 8 and also ref. 31). Accumulation of DNA breaks in the loop anchorage regions may be another reason for the concentration of recombination junctions in these regions.

11.2. VIRAL GENOMES AS MODEL LOOPS

It has long been known that viruses utilize the cellular transcription and replication machinery for replication and transcription of their DNA. Now it has become clear that in order to be able to do that the viral genomes should integrate in a very specific fashion into the preexisting nuclear architecture. Indeed, it has been shown that similarly to transcription of cellular genes, the transcription of viral DNA and subsequent splicing of viral transcripts occur in well-defined nuclear compartments known as foci or speckles.[32] Furthermore, the transport of viral transcripts to the cytoplasm follows distinct pathways which are not affected by the high-salt extraction of nuclei.[33] The positions of replication sites of DNA-containing viruses replicating in nuclei are also determined by the preexisting nuclear architecture, i.e., by the nuclear matrix.[34] These recent data support a number of previous observations indicating that replication and transcription of viral DNA proceed in connection with the nuclear matrix.[35-44] The association of viral genomes with the nuclear matrix is probably mediated by virus-encoded proteins which integrate in the host cell-nuclear matrix. Among these proteins are the key proteins involved in viral DNA replication, such as T-antigen of SV-40 and polyoma viruses[45-51] and terminal proteins of adenoviruses.[52,53] Some other DNA-binding proteins encoded by viral genomes (such as the DNA-binding proteins of herpes simplex virus and Epstein-Barr virus nuclear antigen-2A and C3d receptor protein), which may or may not be involved into the organization of viral DNA replication were also found to reside in the nuclear matrix compartment of eukaryotic cell nuclei.[54-57] The assembly of virions was shown to proceed on the nuclear matrix and hence at late stages of infection both viral capsid proteins and full-length viral DNA molecules can be found in association with the nuclear matrix.[37,43,58,59] In terms of our discussion the most interesting observation is, however, that at least at the early stage of infection replicating viral genomes are bound to the nuclear matrix through replication origins. The same is true of plasmids containing viral replication origins.[39,60-63] Hence viral genomes imitate in a way the chromosomal DNA loop-domains. It is likely that cellular episomal DNA behaves in a similar fashion. Evidence for direct binding of amplified extrachromosomal DNA to the nuclear matrix has been obtained recently.[64]

11.3. EVOLUTION OF THE GENOME THROUGH ACQUIRING AND REPOSITIONING DNA LOOP-DOMAINS

Summarizing the data discussed in the previous two sections we would like to suggest that recombination between loop anchorage regions may constitute a general way of the whole genome evolution involving the acquisition of new domains and the exchange of chromosomal positions of the existing domains. In connection with our suggestion, it is remarkable that in the course of lytic infection the genomes of DNA containing viruses imitate chromosomal DNA loops in a sense that they also become associated with the nuclear matrix through regions containing replication origins. Under certain conditions this association may facilitate the integration of viral genomes into chromosomal DNA (again by illegitimate recombination between loop anchorage sites).

One may argue that the assembly of a complex genome out of simple ones resembling the genomes of modern viruses cannot explain why some loop anchorage sites are located within transcription units. However, long genes covering hundreds of kilobases of DNA may represent a relatively new evolutionary finding. Ribosomal genes are clearly some of the most conservative genes in the eukaryotic genome and in the ribosomal gene cluster the loop anchorage regions are located in nontranscribed spacers. It will be interesting to see how the organization of chromosomal DNA into loops is related to the organization of transcription units in lower eukaryotes. It may happen that in the genome of these organisms the loop anchorage regions are also located strictly outside the transcription units.

Unfortunately, there is no experimental approach to check whether indeed the eukaryotic chromosomes were assembled via recombination between several simple genomes giving rise to the first domains. What, however, seems to be clear is that not only the rearrangement of modern genomes, but also their disintegration in the course of regulated cell death (apoptosis) starts with the accumulation of breaks at the domain borders resulting in the excision of DNA loops and their oligomers (see chapter 8). Hence, it is possible that DNA loops constitute certain basic units of the genome organization.

11.4. CONCLUSIONS

Rearrangements of the genome frequently occur via illegitimate recombination between loop anchorage sites resulting in deletion or repositioning of DNA loops. The illegitimate recombination between loop anchorage sites is possibly mediated by topoisomerase II. Treatments stabilizing intermediate covalent complexes of topoisomerase II with DNA seem to increase the possibility of illegitimate recombination between loop anchorage regions. On the basis of these and some other observations, we suggest that DNA loops constitute structural blocks of the eukaryotic genome.

Related Review Articles

Auxenfants E, Morel P, Lai JL, Sartiaux C, Detourmignies L, Bauters F and Fenaux P. Secondary acute lymphoblastic leukemia with t (4;11): report on two cases and review of the literature. Ann Hematol 1992; 65:143-146.

Ciampor F. The role of cytoskeleton and nuclear matrix in virus replication. Acta Virol (Czechoslovakia) 1988; 32:168-189.

Deppert W, Schirmbeck R. The nuclear matrix and virus function. Int Rev Cytol 1995; 162A:485-537.

Ferguson LR, Baguley BC. Topoisomerase II enzymes and mutagenicity. Environ Mol Mutagen 1994; 24:245-261.

Pui CH. Acute leukemias with the t(4;11) (q21;q23). Leuk Lymphoma 1992; 7:173-179.

Rowley JD. 1993 Robert R. deVilliers Lecture. Chromosome translocations: dangerous liaisons. Leukemia 1994; 8, Suppl. 1:1-6.

References

1. Henglein B, Synovzik H, Groitl P. Three breakpoints of variant t(2;8) translocations in Burkitt's lymphoma cells fall within a region 140 kilobases distal from *c-myc*. Moll Cell Biol 1989; 9:2105-2113.
2. Geng JP, Tong JH, Dong S et al. Localization of the chromosome 15 breakpoints and expression of multiple PML-RAR alpha transcripts in acute promielocytic leukemia: a study of 28 Chinese patients. Leukemia 1993; 7:20-26.
3. Dong S, Geng JP, Tong JH et al. Breakpoint clusters of the PML gene in acute promielocytic leukemia: primary structure of the reciprocal products of the PML-RARA gene in a patient with t(15;17) Genes Chromosom. Cancer 1993; 6:133-139.

4. Sperry AO, Blasquez VB, Garrard WT. Disfunction of chromosomal loop attachment sites: illegitimate recombination linked to matrix associated regions and topoisomerase II. Proc Natl Acad Sci USA 1989; 86:5497-5501.
5. Blasquez VC, Sperry AO, Cockerill PN et al. Protein: DNA interactions at chromosomal loop attachment sites. Genome 1989; 31:503-509.
6. Mita S, Rizzuto R, Moraes CT et al. Recombination via flanking direct repeats is major cause of large-scale deletions of human mitochondrial DNA. Nucl Acids Res 1990; 18:561-567.
7. Chen Z, Chen SJ. RARA and PML genes in acute promyelocytic leukemia. Leuk Lymphoma 1992; 8:253-260.
8. Azou Y, Laval M. Sequence of the novel joints present in the amplified DNA of N-phosphonacetyl-L-aspartate resistant *Drosophila* cells: implication on the mechanisms of amplification in these cells. Biol Cell 1993; 77:155-164.
9. Domer PH, Fakharzadeh SS, Chen CS et al. Acute mixed-leneage leukemia t(4;11)(q21;23) generates an MLL-AF4 fusion product. Proc Natl Acad Sci USA 1993; 90:7884-7888.
10. Blok RB, Thorburn DR, Thompson GN et al. A topoisomerase II cleavage site is associated with a novel mitochondrial DNA deletion. Hum Genet 1995; 95:75-81.
11. Mielke C, Maass K, Tummler M, Bode J. Anatomy of highly expressed chromosomal sites targeted by retroviral vectors. Biochemistry 1996; 35:2239-2252.
12. Gale KC, Osheroff N. Intrinsic intermolecular DNA ligation activity of eukaryotic topoisomerase II. Potential roles in recombination. J Biol Chem 1992; 267:12090-12097.
13. Maraschin J, Dutrillaux B, Aurias A. Chromosome aberrations induced by etoposide (VP-16) are not random. Int J Cancer 1990; 46:808-812.
14. Han Y-H, Finley-Austin MJ, Pommier Y et al. Small deletion and insertion mutantions induced by topoisomerase II inhibitor teniposide in CHO cells and comparison with sites of drug-stimulated cleavage in vitro. J Mol Biol 1993; 229:52-66.
15. Shibuya ML, Ueno AM, Vannais DB et al. Megabase pair deletions in mutant mammalian cells following exposure to amsacrine, an inhibitor of DNA topoisomerase II. Cancer Res 1994; 54:1092-1097.
16. Ratain MJ, Kaminer LS, Bitrain JD et al. Acute nonlymphocytic leukemia following etoposide and cisplatin combination chemotherapy for advanced non-small carcinoma of the lung. Blood 1987; 70:1412-1417.
17. Auxenfants E, Morel P, Lai JL et al. Secondary acute lymphoblastic leukemia with t (4;11): report on two cases and review of the literature. Ann Hematol 1992; 65:143-146.

18. Pui CH. Acute leukemias with the t(4;11) (q21;q23). Leuk Lymphoma 1992; 7:173-179.
19. Super HJ, McCabe NR, Thirman MJ et al. Rearrangements of MLL gene in therapy-related acute myeloid leukemia in patients previously treated with agents targeting DNA-topoisomerase II. Blood 1993; 82:3705-3711.
20. Rowley JD. Rearrangements involving chromosome band 11Q23 in acute leukaemia. Semin Cancer Biol 1993; 4:377-385.
21. Rowley JD. 1993 Robert R. deVilliers Lecture. Chromosome transloctions: dangerous liasions. Leukemia 1994; 8, Suppl. 1:1-6.
22. Bae YS, Chiba M, Ohira M et al. A shuttle vector for analysis of illegitimate recombination in mammalian cells: effects of DNA topoisomerase inhibitors on deletion frequency. Gene 1991; 101:285-289.
23. Bodley AL, Huang HC, Yu C et al. Integration of simian virus 40 into cellular DNA occurs at or near topoisomerase II cleavage hot spots induced by VM-26 (teniposide). Mol Cell Biol 1993; 13:6190-6200.
24. Fernandes DJ, Danks MK, Beck WT. Decreased nuclear matrix DNA topoisomerase II in human leukemia cells resistant to VM-26 and m-AMSA. Biochemistry 1990; 29:4235-4241.
25. Fernandes DJ, Catapano CV. Nuclear matrix targets for anticancer agents. Cancer Cells 1991; 3:134-140.
26. Gromova II, Thomsen B, Razin SV. Different topoisomerase II antitumour drugs direct similar specific long-range fragmentation of an amplified *c-myc* gene locus in living cells and in high-salt-extracted nuclei. Proc Natl Acad Sci USA 1995; 92:102-106.
27. Tsutsui K, Tsutsui K, Oda T. Incorporation of exogenous circular DNA into large catenated networks in isolated nuclei. Evidence for involvement of the nuclear scaffold. J Biol Chem 1989; 264: 7644-7652.
28. Dave VP, Modak MJ, Pandey VN. Nuclear matrix bound V(D)J recombination activity in rat thymus nuclei: an in vitro system. Biochemistry 1991; 30:4763-4767.
29. Pandey VN, Dave VP, Amrute SB et al. Thymic nuclear matrix associated activity is not V(D)J recombinase. Biochem Biophys Res Commun 1991; 181:95-99.
30. Windle B, Draper BW, Yin Y et al. A central role for chromosomal breakage in gene amplification, deletion formation, and amplicone integration. Genes Dev 1991; 5:160-174.
31. Gromova II, Nielsen OF, Razin SV. Long-range fragmentation of the eukaryotic genome by exogenous and endogenous nucleases proceeds in a specific fashion via preferential DNA cleavage at matrix attachment sites. J Biol Chem 1995; 270:18685-18690.
32. Jimenez-Garsia LF, Spector DL. In vivo evidence that transcription and splicing are coordinated by a recruiting mechanism. Cell 1993; 73:47-59.

33. Xing YG, Lawrence JB. Preservation of specific RNA distribution within the chromatin-depleted nuclear substructure demonstrated by in situ hybridization coupled with biochemical fractionation. J Cell Biol 1991; 112:1055-1063.
34. De Bruyn Kops A, Knipe DM. Preexisting nuclear architecture defines the intranuclear location of herpesvirus DNA replication structures. J Virol 1994; 68:3512-3526.
35. Ben-Ze'ev A, Abulafia R, Alony Y. SV40 virions and viral RNA metabolism are associated with cellular structures. EMBO J 1982; 1:1225-1231.
36. Abulafia R, Ben-Ze'ev A, Hay N et al. Control of late simian virus 40 transcription by the attenuation mechanism and transcriptionally active ternary complexes are associated with the nuclear matrix. J Mol Biol 1984; 172:467-487.
37. Smith HC, Berezney R, Brewster JM et al. Properties of adenoviral DNA bound to the nuclear matrix. Biochemistry 1985; 24: 1197-1202.
38. Younghusband HB. An association between replicating adenovirus DNA and the nuclear matrix in infected HeLa cells. Can J Biochem Cell Biol 1985; 63:654-660.
39. Jones C, Su RT. Association of viral and plasmid DNA with the nuclear matrix during productive infection. Biochim Biophys Acta 1987; 910:52-62.
40. Watson JB, Gralla JD. Simian virus 40 associates with nuclear superstructures at early times of infection. J Virol 1987; 61:748-754.
41. Zhai ZH, Chen F, Nickerson JA et al. Association of adenovirus DNA transcribed activity with nuclear matrix of host cells. Sci China (China) 1989; 32:176-185.
42. Schirmbeck R, Deppert W. Structural topography of simian virus 40 DNA replication. J Virol 1991; 65:2578-2588.
43. Schirmbeck R, Von der Weth A, Deppert W. Structural requirements for simian virus 40 replication and virion maturation. J Virol 1993; 67:894-901.
44. Besse S, Puvion-Dutilleul F. Compartmentalization of cellular and viral DNAs in adenovirus type 5 infection as revealed by ultrastructural in situ hybridization. Chromosome Res 1994; 2:123-135.
45. Buckler-White AJ, Humphrey GW, Pigiet V. Association of polyoma T antigen and DNA with the nuclear matrix from lytically infected 3T6 cells. Cell 1980; 22:37-46
46. Verderame MF, Kohtz DS, Pollack RF. 94,000- and 100,000-molecular-weight simian virus 40 T-antigens are associated with the nuclear matrix in transformed and revertant mouse cells. J Virol 1983; 46:575-583.
47. Jones C, Su RT. Association of phosphorilated simian virus 40 T-antigen with subnuclear fractions of infected and transformed cells. Exp Cell Res 1985; 160:158-170.

48. Schirmbeck R, Deppert W. Specific interaction of simian virus 40 large T antigen with cellular chromatin and nuclear matrix during the course of infection. J Virol 1987; 61:3561-3569.
49. Schirmbeck R, Deppert W. Nuclear subcompartmentalization of simian virus 40 large T antigen: evidence for in vivo regulation of biochemical activities. J Virol 1989; 63:2308-2316.
50. Puvion E, Duthu A, Harper F et al. Intranuclear distribution of SV40 large T-antigen and transformation-related protein p53 in abortively infected cells. Exp Cell Res 1988; 177:73-89.
51. Mann K. Characterization of SV40 T antigen associated with the nuclear matrix. Virology 1990; 179:437-445.
52. Schaack J, Ho WY, Freimuth P et al. Adenovirus terminal protein mediates both nuclear matrix association and efficient transcription of adenovirus DNA. Genes Dev 1990; 4:1197-1208.
53. Fredman JN, Engler JA. Adenovirus precursor terminal protein interacts with the nuclear matrix in vivo and in vitro. J Virol 1993; 67:3384-3395.
54. Pinard MF, Simard R, Bibor-Hardy V. DNA-binding proteins of herpes simplex virus type 1-infected BHK cell nuclear matrices. J Gen Virol 1987; 68:727-735.
55. Burdett LA, Docherty JJ, Howett MK. Nucleocapsid, nuclear association, and genome location of the herpes simplex virus type 2 38-kD DNA binding protein. Intervirology 1990; 31:76-84.
56. Grasser FA, Haiss P, Gottel S et al. Biochemical characterization of Epstein-Barr virus nuclear antigen 2A. J Virol 1991; 65: 3779-3788.
57. Chang YE, Roizman B. The product of the UL31 gene of herpes simplex virus 1 is a nuclear phosphoprotein which partitions with the nuclear matrix. J Virol 1993; 67:6348-6356.
58. Bibor-Hardy V, Dagenais A, Simard R. In situ localization of the major capsid protein during lytic infection by herpes simplex virus. J Gen Virol 1985; 66: 897-901.
59. Khittoo G, Delorme L, Dery CV et al. Role of the nuclear matrix in adenovirus maturation. Virus Res 1986; 4:391-403.
60. Bodnar JW, Hanson PI, Polvino-Bodnar M et al. The terminal regions of adenovirus and minute virus of mice DNAs are preferentially associated with the nuclear matrix in infected cells. J Virol 1989; 63:4344-4353.
61. Adom JN, Richard-Foy H. A region immediately adjacent to the origin of replication of bovine papilloma virus type 1 interacts in vitro with the nuclear matrix. Biochem Biophys Res Commun 1991; 176:479-485.
62. Adom JN, Gouilleux F, Richard-Foy H. Ineraction with the nuclear matrix a chimeric construct containing a replication origin and a transcription unit. Biochim Biophys Acta 1992. 1171:187-197.

63. Jankelevich S, Kilman JL, Bodnar JW et al. A nuclear matrix attachment region organizes the Epstein-Barr viral plasmid in Raji cells into a single DNA domain. EMBO J 1992; 11:1165-1176.
64. Thiebaut F, Hanauske AR, Von Hoff DD. Evidence for binding of extrachromosomal DNA sequences to nuclear matrix proteins in multidrug-resistant KB-V1 cells. FEBS Lett 1993; 319:133-137.

CHAPTER 12

Models Describing the Nuclear Matrix Organization and Compartmentalization of the Nuclear Space

12.1. THE CHANNEL MODEL OF THE NUCLEAR MATRIX

At first glance there seems to be a contradiction between the localization of DNA loop anchorage regions on the nuclear matrix and their preferential accessibility to nucleases. Indeed, the question arises as to how the same structure (nuclear matrix) may function as a scaffolding element for the assembly of chromosomal DNA loops and simultaneously as a place where active functional processes occur. It could possibly be expected that the loop basements are hidden somewhere within a chromosome and hence are not accessible to nucleases. Similarly, it seems difficult to reconcile the data on preferential sensitivity of active genes to nucleases[1] with the data on their attachment to the nuclear matrix (see chapters 4, 7). Indeed, the latter suggest that active genes should be solubilized from nuclei after nonfunctioning DNA sequences located in the loops, while the former demonstrates that treatment of permeabilized cells or isolated nuclei with nucleases solubilizes active genes in the first turn. The reversed pattern of solubilization of transcriptionally active DNA sequences was observed only after chromatin disruption by high-salt extraction (see chapter 4).

Recently, we have proposed a new nuclear matrix model[2] which seems to resolve the above contradiction as well as some other intrinsic problems of the whole "nuclear matrix story." The main assumption constituting the basis of the above model is that the scaffolding element providing a surface for the attachment of DNA loops is not a filament but a channel connected with a nuclear pore. Chromosomal DNA domains are thought to be assembled around this channel, as shown in Figure 12.1. The internal nuclear matrix is hence postulated to be a network of such channels (or pathways) which provide possibilities for the ongoing transport of different compounds (precursors for DNA and RNA synthesis, enzymes and regulatory proteins) from the cytoplasm and for the opposite transport of messenger RNA from the sites of synthesis and splicing to the cytoplasm. Therefore the channel model of nuclear matrix may be considered as further development of the gating hypothesis proposed by Blobel.[3] The channels are not considered to be a uniform-sized structure. On the contrary, it seems likely that they may branch and locally expand giving rise to caverns some of which may provide a surface for the assembly of replicon clusters,[4,5] while the others may correspond to the sites of pre-mRNA splicing.[6-8] It is important that replication foci and splicing speckles have been observed both in nonextracted nuclei and in nuclear matrices indicating that their integrity is not affected by high-salt extraction.[4,9,10]

The conventional model of nuclear matrix (filamentous network) explains associations of functionally active DNA sequences with the nuclear matrix by the necessity to organize spatially multienzyme complexes accomplishing replication and transcription.[11] In addition to this function of the nuclear matrix, the channel model postulates that the attachment of functionally active DNA sequences to the nuclear matrix renders them preferentially accessible for trans-acting factors. The suggestion is corroborated by a recent observation indicating that DNA precursors are channeled directly to the places of DNA replication (replication foci) on the nuclear matrix.[12] Nopp140, a protein which recognizes the nuclear localization signals and shuttles between the cytoplasm and the nucleolus (possibly organizing the transport of ribosomal proteins

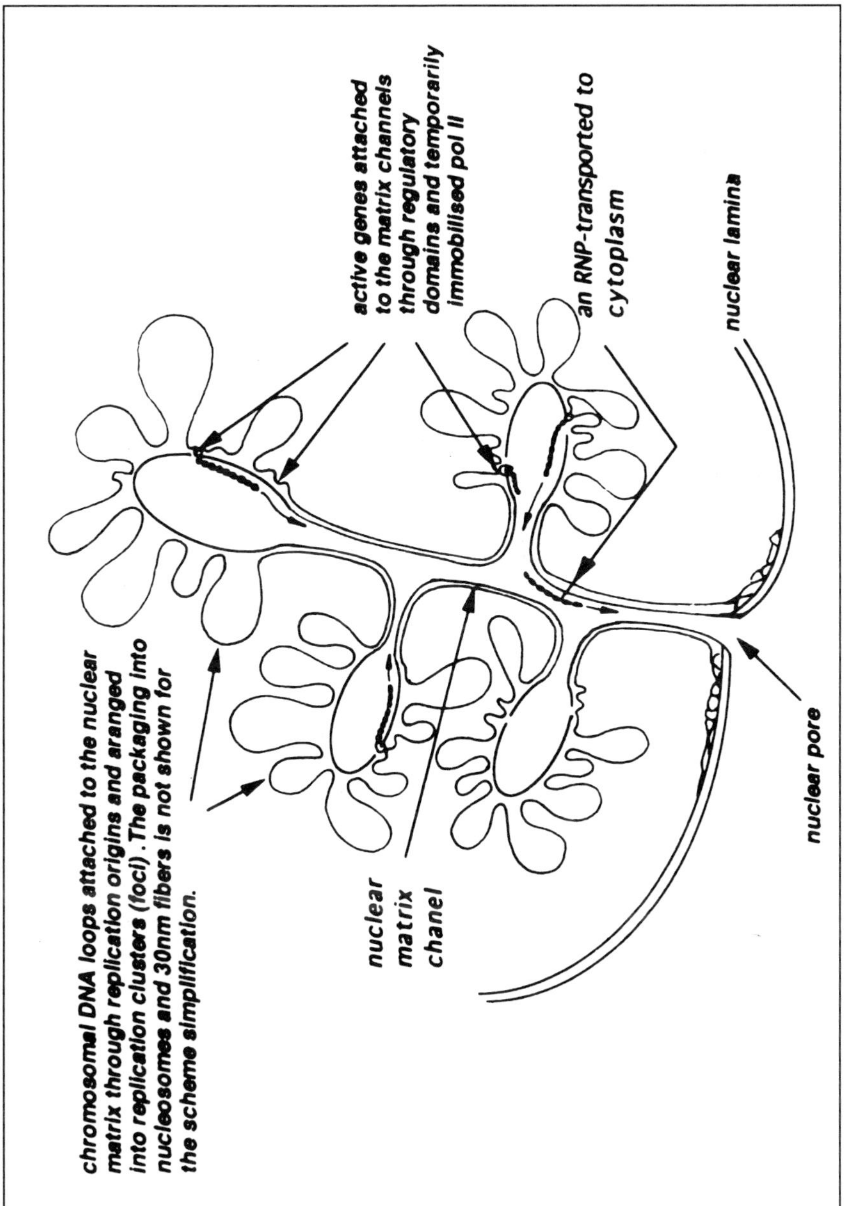

Fig. 12.1. The model of an interphase chromosome organization based on an assumption that the nuclear matrix is a system of internal nuclear channels. (Reprinted with permission of ICSU Press from Razin SV, Gromova II. BioEssays 1995; 17: 443-450. © 1995 ICSU Press).

into nucleolus), was found to move along distinct pathways (channels?). These pathways can be visualized by immunostaining as tracks starting from the dense fibrillar component of the nucleolus and extending to a subset of nuclear pore complexes.[13] The channel model explains why not only DNA sequences actually involved in functional processes, but also regulatory elements (replication origins, enhancers, etc.) are located in proximity to the nuclear matrix. According to the model, this makes these sequences preferentially accessible for trans-acting factors. The model also explains why DNA sequences attached to the nuclear matrix are preferentially sensitive to nucleases. Indeed, it may happen that under certain experimental conditions exogenous nucleases will penetrate into matrix channels and hence preferentially digest exactly those DNA sequences that are bound to the nuclear matrix. Both replicating and transcribing DNA sequences temporarily associated with the nuclear matrix and the DNA loop anchorage sites will constitute preferential targets for the nucleases. Similarly, at the initial steps of apoptosis the matrix channels may serve as pathways directing cytoplasmic nucleases to the DNA loop anchorage sites.

High-salt extraction and some other treatments used for histone removal may possibly cause a collapse of the matrix channels and the aggregation of the internal staff. This may explain the reversed pattern of nuclease sensitivity of matrix-bound DNA sequences in high-salt-extracted nuclei. After the collapse of the matrix channels the chromosomal DNA will be accessible only from the outside. Theis means that the matrix attached DNA sequences will become partially protected and that their solubilization in the course of nuclease digestion will occur in the last turn, as was indeed observed in a number of experiments with high-salt-extracted nuclei (for a review see refs. 11, 14, 15).

Under natural conditions the transport of mRNA to the cytoplasm is likely to be one of the main functions of the nuclear matrix channels. It has been indeed shown that RNA synthesis, processing and transport are spatially organized along some distinct pathways within nuclei[7,8,16-18] (for a review see refs. 19, 20). At least in some cases these pathways (channels?) could be traced up to the nuclear envelope.[18] Furthermore, the specific distribution of individual transcripts and at least some splicing factors is not affected by high-salt extraction,[9,21] indicating that the above pathways (channels?) constitute a part of the nuclear matrix. Once the RNA transport is blocked for any reason, the channels will be overloaded with RNP particles that can easily aggregate[22,23] to form RNA-containing filaments inside the channel. In this way RNP-containing filamentous networks[24] can be created.

It is necessary to stress that the channel model of nuclear matrix incorporates all basic ideas of the conventional conception of nuclear matrix (i.e., that the matrix supports the nuclear architecture, provides surfaces for the assembly of different enzyme complexes and also serves for compartmentalization of the nuclear space). In addition, the channel model postulates the involvement of the nuclear matrix into the organization of macromolecular transport within nuclei. The keypoint of the model is that DNA sequences bound to the nuclear matrix are preferentially accessible to soluble factors (including regulatory proteins and different soluble enzymes). This assumption helps to explain the reverse pattern of nuclease sensitivity of active genes in permeabilized cells and in high-salt-extracted nuclei. It also provides additional rationales for the accomplishment of replication and transcription on the nuclear matrix. Artificial creation of RNP-containing filamentous internuclear networks in the course of isolation of nuclear matrices can easily be explained by the suggested model. Finally, the model proposes how different protein factors may be directly transported from the places of their synthesis in cytoplasm to specific target sequences on chromosomal DNA. The operation of some regulatory mechanisms choosing the necessary targets by a simple selection of channels seems to be quite possible at this level. An advantage of such a mechanism would be that it can ensure simultaneous activation of a selected group of independent targets connected with the same channel, for example all replicons in a given cluster.

12.2. CHROMOSOME TERRITORIES AND THE INTERCHROMOSOME DOMAIN COMPARTMENT (ICD) MODEL OF ORGANIZATION OF EUKARYOTIC CELL NUCLEI

Although during mitosis individual chromosomes represent separate entities, it does not mean that they should behave similarly during interphase. It may be probable that after decondensation different chromosomes invade each other or even participate in mutual organization of a complex three-dimensional chromatin network. One may, for example, propose that active genes of different chromosomes are concentrated in the same nuclear domain. The first experimental analysis of mutual distribution of chromosomes in interphase nuclei was carried out by Cremer and collaborators.[25-28] These authors introduced a local damage into the DNA of interphase nuclei by microbeam UV irradiation and then allowed the cells to repair the above damage using radiolabeled precursors or nucleotide analogs. After entering metaphase, the cells were fixed and the repaired areas were visualized by autoradiography or immunostaining. The results demonstrate that each area of interphase nuclei contains parts of only a few chromosomes. In opposite experiments (when metaphase chromosomes were microirradiated), the damaged regions occupied a relatively compact area after transition from the metaphase chromosomes to the interphase nuclei. A conclusion was drawn that in interphase nuclei individual chromosomes occupy distinct territories.[25-28]

Generation of species-specific and chromosome-specific sets of DNA probes and the development of the technique of suppressive hybridization open up possibilities for analyzing more precisely the distribution of individual chromosomes in eukaryotic cell nuclei by fluorescence in situ hybridization. The results of these studies support the idea of territorial organization of interphase chromosomes. Different chromosomes were found to occupy relatively large, but limited and nonoverlapping domains within nuclei.[29-34] These domains are usually referred to as "chromosome territories." Homologous chromosomes were not necessarily found close to each other.[35] In general, no strict positioning of chromosomes was detected. Pairs of chromosomes occupying neighboring territories in one nucleus can be widely separated in another one. Although it is a common belief that transcriptionally active genomic domains are locally unfolded, this unfolding turned out to be rather limited. Indeed, the volume of an active copy of X-chromosome in human female cells was found to exceed the volume of its heterochromatic homologue by 25% only.[35,36] Most interestingly, active genes were found exclusively at the periphery of the chromosome territories[37,38] although not necessarily at the nuclear periphery, as suggested by some previous observations.[39] Furthermore, RNA tracks and splicing components also were found at the periphery of chromosome territories or even outside them.[38,40] To explain all these findings, the transcription and splicing of pre-mRNA were proposed to occur in the Interchromosome Domain Compartment (ICD). According to the suggested model,[37,40,41] this compartment was considered to be a space between the adjacent chromosome territories (Fig. 12.2). Transport of mRNA from the places of transcription and splicing to the nuclear periphery was also proposed to proceed within the ICD compartment or, in other words, outside the chromosome territories. Finally, it was proposed that the existence of an ICD compartment is granted by repulsive Debye-Huckel forces between the surfaces of similarly charged neighboring chromosome territories.

An obvious advantage of the ICD compartment model is that it is based on structural studies demonstrating that active genes, splicing components and hnRNA are located at the periphery of the chromosome territory. The above observations are further supported by the results of Zachar and collaborators.[42] In this work the transport of transcripts of a specific gene expressed in polythene nuclei of *Drosophila* silver glands was studied. Evidence was presented that splicing and transport of the above transcripts occurred in a so-called "extrachromosomal network" defined by the exclusion from nucleoli and chromosome axes (i.e., in the ICD compartment). However, maintaining of the ICD compartment by simple electrical forces can be questioned. Indeed, as has been discussed above, the splicing speckles and RNA tracks are preserved in nuclear matrices after digestion of chromosomal DNA and high-salt extraction. These observations argue for the existence of some solid structure (nuclear matrix) which supports the nuclear architecture even when the repulsive forces between chromosomal territories are eliminated. One may propose that at least some high salt-resistant nuclear structures are created as a result of aggregation of RNP particles within ICD compartment. However, the possibility to observe internal nuclear matrix in nuclei depleted of chromatin by means of electroelution in a buffer with physiological ionic strength[43,44] strongly suggest that a nonchromatin skeletal structure of some kind preexist in living cells.

Although apparently different, the channel model of nuclear matrix and the ICD compartment model are not incompatible with each other. Indeed, the channel model represents a completely logical creation and it does not predict whether channels span the chromosome or cover it from outside. In the latter case, the ma-

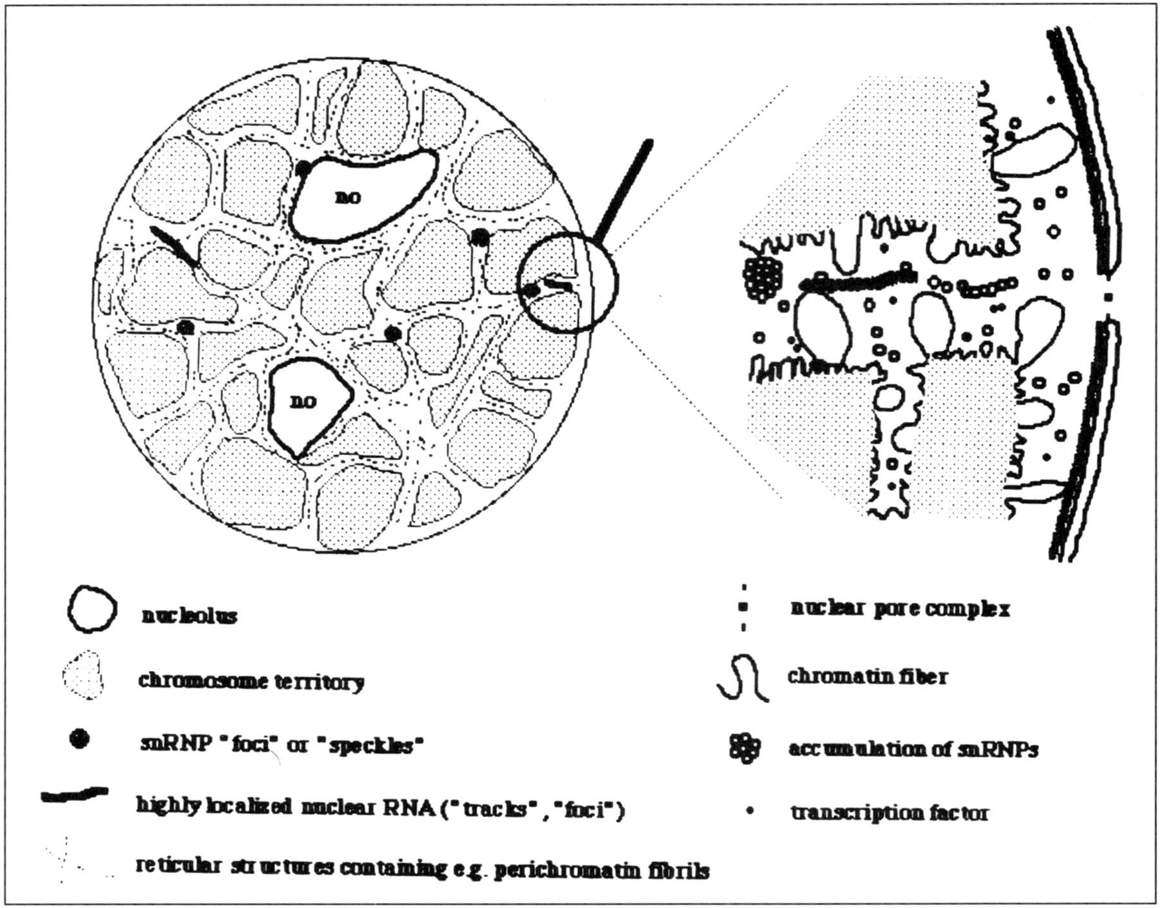

Fig. 12.2. Schematic drawing of an optical section through a cell nucleus according to the ICD-compartment model[40] of a functional nuclear compartmentalization. A section through a whole nucleus (left) and a detailed view at higher magnification (right) are shown. The area between the chromosome territories constitutes the ICD compartment. Partial DNA loops extending into the ICD space are shown on the right. According to the model, active genes occur at the surface of the territories, and RNA transcripts are directly released into ICD space. Transcription factors are mostly restricted to the ICD compartment, greatly reducing the volume for searching DNA target sequences. Splicing components are concentrated in the ICD space as well. The reticular ICD space also contains nuclear structures enriched in splicing factors like the perichromatin fibrils, interchromatin granules, and coiled bodies. The chromosome territories are nonoverlapping but may attach at some sites to the nuclear envelope or with each other. The ICD compartment extends between the sites of attachment and is associated with the nuclear pore complexes. Note that in an optical section, the ICD space may appear as channels. The surface of chromosome territories may be greatly increased by infoldings, as outlined within the magnified section. (Reprinted with permission of Cold Spring Harbor Laboratory Press from Cremer T, Kurz A, Zirbel R et al. Cold Sprig Harb Symp Quant Biol 1993; 58:777-792. © 1993 Cold Spring Harbor Laboratory Press.)

trix channels will strongly resemble the ICD compartment, suggesting that this compartment is not simply maintained by repulsive electrostatic forces but possesses some skeletal elements (such as nuclear matrix filaments).

Whatever the nature of nuclear matrix channels may be, the transport of mRNA along these channels may proceed in either active or passive fashion. At present there is no direct experimental evidence permitting one of the above possibilities to be chosen. Some indirect data are more in agreement with the supposition that RNA is not simply diffused along the channels. Indeed, RNA cannot be solubilized by the extraction of nuclei with either a low salt or a high salt solution. However, this solubilization can be achieved by treatment of nuclear matrices with cytochalasin B, an agent causing depolymerization of actin filaments,[45,46] indicating that during splicing and transport the mRNA remains temporarily attached to the nuclear skeleton. The supposition that RNA cannot be extracted from nuclear matrices due to the aggregation of RNP proteins and formation of RNA-containing filaments cannot explain the effect of cytochalasin. In cells infected by adenoviruses, the transport to the cytoplasm of host-cell mRNAs is suppressed while viral transcripts easily reach the cytoplasm.[47] The selection of transcripts which are allowed to leave nuclei is made by a complex of virus-encoded proteins (E1B-E4) which is not associated with the nuclear pores or the nuclear envelope but is located within nuclei. It is again easier to explain these data assuming that the transport of RNA along the nuclear matrix channels is an active process.

12.3. CONCLUSIONS

Besides serving as a skeletal structure supporting the nuclear architecture, the nuclear matrix is likely to organize pathways for macromolecular transport within nuclei. These pathways or channels may be accessible to exogenous enzymes in permeabilized cells, but not in high-salt-extracted nuclei. If this supposition is correct, the preferential targets for nuclease cleavage are to be different in permeabilized cells and in nuclear matrices, as has indeed been shown in a number of experiments. The organization of nuclear channels remains to be characterized. Although it was suggested that these channels might be supported by repulsive electric forces between similarly charged chromosomal territories, the situation seems to be more complicated, as such channels (pathways of mRNA transport) can be traced in isolated nuclear matrices and thus are supported by some nonchromatin structure.

Related Review Articles

Haaf T, Schmid M. Chromosome topology in mammalian interphase nuclei. Exp Cell Res 1991; 192:325-332.

Spector DL. Nuclear organization and pre-mRNA processing. Curr Opin Cell Biol 1993; 5:442-447.

Spector DL. Macronuclear domains within the cell nucleus. Annu Rev Cell Biol 1993; 9:265-315.

Cremer T, Kurz A, Zirbel R et al. Role of chromosome territories in the functional compartmentalization of the cell nucleus. Cold Spring Harb Symp Quant Biol 1993; 58:777-792.

Kramer J, Lachar L, Bingham PM. Nuclear pre-mRNA metabolism: channels and tracks. Trend Cell Biol 1994; 4:35-37.

Razin SV, Gromova II. The channels model of nuclear matrix structure. Bioessays 1995; 17:443-449.

References

1. Weintraub H, Groudine M. Chromosomal subunits in active genes have an altered conformation. Science 1976; 73:848-856.
2. Razin SV, Gromova II. The channels model of the nuclear matrix structure. Bioessays 1995; 17:443-450.
3. Blobel G. Gene gating: A hypothesis. Proc Natl Acad Sci USA 1985; 82:8527-8529.
4. Nakayasu H, Berezney R. Mapping replication sites in eukaryotic cell nucleus. J Cell Biol 1989; 108:1-11.
5. Hozak P, Hassan AB, Jackson DA et al. Visualization of replication factories attached to the nucleoskeleton. Cell 1993; 73:361-373.
6. Jackson DA, Hassan AB, Errington RJ et al. Visualization of focal sites of transcription within human nuclei. EMBO J 1993; 12:1059-1065.
7. Xing Y, Johnson CV, Dobner PR et al. Higher level organization of individual gene transcription and RNA splicing. Science 1993; 259:1326-1330.
8. Carter CC, Bowman D, Carrington W et al. A three-dimensional view of precursor messenger RNA metabolism within mammalian nucleus. Science 1993; 259:1330-1335.
9. Smith HC, Ochs RL, Fernandez EA et al. Macromolecular domains containing nuclear protein p107 and U-snRNP protein p28: further evidence for the in situ nuclear matrix. Mol Cell Biochem 1986; 70:151-158.

10. Huang S, Deerinck TJ, Ellisman MH et al. In vivo analysis of the stability and transport of nuclear poly (A)+ RNA. J Cell Biol 1994; 126:877-899.
11. Jackson DA, Dolle A, Robertson G et al. The attachments of chromatin loops to the nucleoskeleton. Cell Biol Int Rep 1992; 16:687-696.
12. Panzeter PL, Ringer DP. DNA precursors are channelled to nuclear matrix DNA replication sites. Biochem J 1993; 293:775-779.
13. Meier UT, Blobel G. Nopp140 shuttles on tracks between nucleolus and cytoplasm. Cell 1992; 70:127-138.
14. Razin SV. DNA interactions with the nuclear matrix and spatial organization of replication and transcription. Bioessays 1987; 6:19-23.
15. Razin SV, Vassetzky YS. Domain organization of eukaryotic genome. Cell Biol Int Rep 1992; 16:697-708.
16. Lawrence JB, Singer RH, Marselle LM. Highly localized tracks of specific transcripts within interphase nuclei visualized by in situ hybridization. Cell 1989; 57:493-502.
17. Lawrence JB, Carter KS Xing X. Probing functional organization within the nucleus: is genome structure integrated with RNA metabolism. Cold Spring Harb Symp Quant Biol 1993; 58:807-818.
18. Spector DL. Higher order nuclear organization: Three-dimensional distribution of small nuclear ribonucleoprotein particles. Proc Natl Acad Sci USA 1990; 87:147-151.
19. Rosbash M, Singer RH. RNA travel: Tracks from DNA to cytoplasm. Cell 1993; 75:399-401.
20. Spector DL. Macromolecular domains within the cell nucleus. Annu Rev Cell Biol 1993; 9:265-315.
21. Xing YG, Lawrence JB. Preservation of specific RNA distribution within the chromatin-depleted nuclear substructure demonstrated by in situ hybridization coupled with biochemical fractionation. J Cell Biol 1991; 112:1055-1063.
22. Akenstorf HP, Convay GC, Wooley JC et al. Nuclear matrix-like filaments form through artifactual rearrangement of hnRNP proteins. J Cell Biol 1984; 99 (4), Pt.2:233.
23. Kloetzel P-M, Johnson RM, Sommerville J. Interaction of the hnRNA of amphibian oocytes with fibril-forming proteins. Eur J Biochem 1982; 127:301-308.
24. Faiferman I, Pogo AO. Isolation of a nuclear ribonucleoprotein network that contains heterogenous RNA and is bound to the nuclear envelope. Biochemistry 1975; 14:3808-3816.
25. Cremer T, Cremer C, Baumann H et al. Rabl's model of the interphase chromosome arrangement tested in Chinese hamster cells by premature chromosome condensation and laser-UV-microbeam experiments. Hum Genet 1982; 60:46-56.
26. Cremer T, Cremer C, Schneider T et al. Analysis of chromosome positions in the interphase nucleus of Chinese hamster cells by la-

ser-UV-microirradiation experiments. Hum Genet 1982; 62: 201-209.

27. Hens L, Baumann M, Cremer T et al. Immunocytochemical localization of chromatin regions UV-microirradiated in S phase or anaphase. Exp Cell Res 1983; 149:257-269.
28. Cremer T, Baumann H, Nakanishi K et al. Correlation between interphase and metaphase chromosome arrangements as studied by laser-UV-microbeam experiments. Chromosomes today 1984; 8:203-206.
29. Schardin M, Hager HD, Lang M. Specific staining of human chromosomes in chinese hamster x man hybrid cell lines demonstrates interphase chromosome territories. Human Genet 1985; 71:281-287.
30. Manuelidis L. Individual interphase domains revealed by in situ hybridization. Human Genet 1985; 71:288-293.
31. Van Dekken H, Pinkel D, Mullikin J et al. Three-dimensional analysis of the organization of human chromosome domains in human and human-hamster hybrid interphase nuclei. J Cell Sci 1989; 94:299-306.
32. Cremer T, Lichter P, Borden J et al. Detection of chromosome aberrations in metaphase and interphase tumor cells by in situ hybridization using chromosome specific library probes. Hum Genet 1988; 80:235-246.
33. Lichter P, Cremer T, Borden J et al. Delineation of individual human chromosomes in metaphase and interphase cells by in situ suppression hybridization using recombinant DNA libraries. Hum Genet 1988; 80:224-234.
34. Pinkel D, Landegent J, Collins C et al. Fluorescence in situ hybridization with human chromosome-specific libraries: Detection of trisomy 21 and translocations of chromosome 4. Proc Natl Acad Sci USA 1988; 85:9138-9142.
35. Bischoff A, Albers J, Kharboush I et al. Differences in size and shape of active and inactive X-chromosome domains in human amniotic fluid cell nuclei. Microsc Res Tech 1993; 25:68-77.
36. Eils R, Bertin E, Saracoglu K et al. Application of confocal laser microscopy and three-dimensional Voronoi diagrams for volume and surfase estimates of interphase chromosomes. J Microsc 1995; 177:150-161.
37. Cremer T, Kurz A, Zirbel R et al. Role of chromosome territories in the functional compartmentalization of the cell nucleus. Cold Spring Harb Symp Quant Biol 1993; 58:777-792.
38. Moen PT, Smith KP, Laerence JB. Compartmentalization of specific pre-mRNA metabolism: an emerging view. Hum Mol Genet 1995; 4:1779-1789.
39. Hutshison N, Weintraub H. Localozation of DNase 1-sensitive sequences to specific regions of interphase nuclei. Cell 1985; 43:471-482.

40. Zibrel RM, Mathieu UR, Kurz A et al. Evidence for a nuclear compartment of transcription and splicing located at chromosome domain boundaries. Chromosome Res 1993; 1:93-106.
41. Cremer T, Dietzel S, Eils R et al. Chromosome territories, nuclear matrix filaments and inter-chromatin channels: a topological view on nuclear architecture and function. In: Brandham PE, Bennett MD, eds. Kew Chromosome Conference IV: Royal Botanic Gardens, Kew, 1995:63-81.
42. Zachar Z, Kramer J, Mims IP et al. Evidence for channelled diffusion of pre-mRNAs during nuclear RNA transport in metazoans. J Cell Biol 1993; 121:729-742.
43. Jackson DA, Cook PR. Visualization of a filamentous nucleoskeleton with a 23 nm axial repeat. EMBO J 1988; 7:3667-3677.
44. Hozak P, Hassan AB, Jackson DA et al. Visualization of replication factories attached to the nucleoskeleton. Cell 1993; 73:361-373.
45. Nakayasu H, Ueda K. Association of rapidly-labeled RNAs with actin in nuclear matrix from mouse L5178Y cells. Exp Cell Res 1985; 160:319-330.
46. Schroder HC, Trolltsch D, Wenger R et al. Cytocholasin B selectively releases ovalbumin mRNA precursors but not the mature ovalbumin mRNA from hen oviduct nuclear matrix. Eur J Biochem 1987. 167:239-245.
47. Leppard KN, Shenk T. The adenovirus E1B 55 kd protein influences mRNA transport via an intranuclear effect on RNA metabolism. EMBO J 1989; 8:2329-2336.

INDEX

U

V

X